Strategic Continuous
Process Improvement

About the Author

Gerhard Plenert, Ph.D., CPIM, is a practice partner at Wipro Consulting and president of the Institute of World Class Management. He has over 25 years of professional experience in IT quality and productivity consulting and in manufacturing planning and scheduling methods. Dr. Plenert has worked in the private sector for both corporations and consultants; in the government sector on the federal, state, local, and international levels; and in the academic sector as a college faculty member. He earned a doctorate in mineral economics at the Colorado School of Mines, and has published 10 previous books and over 150 articles.

Strategic Continuous Process Improvement

Which Quality Tools to Use,
and When to Use Them

Gerhard Plenert, Ph.D., CPIM

New York   Chicago   San Francisco
Lisbon   London   Madrid   Mexico City
Milan   New Delhi   San Juan
Seoul   Singapore   Sydney   Toronto

McGraw-Hill books are available at special quantity discounts to use as premiums and sales promotions, or for use in corporate training programs. To contact a representative, please e-mail us at bulksales@mcgraw-hill.com.

Strategic Continuous Process Improvement

1 2 3 4 5 6 7 8 9 0 QFR/QFR 1 7 6 5 4 3 2 1

ISBN 978-0-07-176718-7
MHID 0-07-176718-5

This book is printed on acid-free paper.

Sponsoring Editor	**Project Manager**	**Indexer**
Judy Bass	Neha Rathor, Neuetype	Robert Swanson
Editing Supervisor	**Copy Editor**	**Art Director, Cover**
Stephen M. Smith	Lisa McCoy	Jeff Weeks
Production Supervisor	**Proofreader**	**Composition**
Pamela A. Pelton	Carol Shields	Neuetype
Acquisitions Coordinator		
Bridget L. Thoreson		

To the love of my life,
Renee Sangray Plenert,
who brings "continuous improvement" to my life!

To my parents, George and Ida Plenert,
who tried their best to introduce quality into me
(not sure what happened there)!

And to my kids, their spouses, and of course the grandkids—
Heidi, Dawn, Gregory & Debbie, Gerick, Joshua & Amy, Natasha &
Mark, Zackary, Chelsey, Lucas, Boston, Evan, Lincoln, and Livy Jay—
who keep my life in Quality Chaos!

CONTENTS

PART I
CPI Framework

PART II
CPI Options

PART III
Executing the Improvement Process

PART IV
World Class CPI

PREFACE

Many years ago, when the United States referred to Japanese products as "Japanese junk," Ford of Canada put a car part out for bid. They had numerous offers, but the winning bid came from Japan. In the contract that was issued, Ford specified that there could not be any more than 5 percent bad parts. Something was lost in the translation and, when Ford received the parts and opened the box, they found a bag of parts at the top of the box and with the bag was a note that read, "We don't understand why you wanted to receive 5 percent bad parts, but we separated them out for you so you would not have any trouble finding them." This event, and many others like it, became the defining line between what was previously "Japanese junk" to what became the world standard for quality. We now imitate the Japanese quality methodology worldwide as the standard for how quality systems should operate. Japanese quality systems have now become the world standard.

The United States and Europe have extended an enormous amount of effort in studying and imitating the Japanese methodology. Within Japan, the quality standard has become the Toyota Production System (TPS) methodology. As the West studied the TPS process, we frequently became focused on some small piece or element of it. First we pulled out quality circles, thinking that was the cure-all for Western quality woes. That didn't seem to solve all the problems. Then we tried statistical process control (SPC) and statistical quality control (SQC). After that we migrated to total quality methodology (TQM). From there we went to Just-in-Time (JIT), Lean, and Six Sigma, and on and on in a never-ending attempt to carve up the TPS process into little pieces, always hoping to identify the magical chunk that explains how this quality system could be so successful. Unfortunately, during all this dissection, we failed to realize that the magic behind the success of TPS is not in the individual tools, but in their cultural integration. The key is having all the tools available and using them appropriately. And that brings us back to the purpose of this book. This book attempts to take all these carved-up pieces and join them together into a large bag of tricks that allows us to optimize our quality environment and not just focus on one isolated quality problem.

But, at a higher level, we find an enormous amount of confusion in the basic question, "What is quality?" If it were possible to come up with one standard definition that could satisfy everyone's expectations, then defining "optimal quality" would be easy. There are entire organizations that pursue a meaningful definition of quality. For example, the *Quality Digest* website defines quality as follows:

> Quality is the ongoing process of building and sustaining relationships by assessing, anticipating, and fulfilling stated and implied needs.
>
> QUALITY DIGEST WEB SITE (www.qualitydigest.com/html/qualitydef.html)

This definition is one of the best. Other definitions seem to go downhill from here. Unfortunately, textbook definitions like this one try to search for the generic definition that covers all situations and thereby creates definitions at too high a level to give us anything meaningful or executable. Identifying a practical definition of quality is a nightmare. When surveying quality systems users for their personal definitions, we end up with the following:

1. Whatever the customer wants it to be
2. Satisfying customer expectations
3. Defect-free product
4. Zero waste
5. No returns
6. And so on

There is no consistency in the definition of quality. One individual's definition is extremely different from another's. It always depends on their perspective: user, customer, supplier, government, and so on.

Defining quality becomes even more complicated because of the many "quality improvement systems" that exist. Each claims to be the key to achieving world class quality. But each defines quality in a way that supports their understanding and application of it. The result is Quality Chaos.

This book is not going to attempt to define quality or to tell you that one quality improvement system is the cure-all for all your quality ills. What this book *will* do is link your quality problem (your definition of quality) with one or more of the many quality improvement systems focused on solving that particular problem. This book focuses on resolving the Quality Chaos by creating a link between quality problems and their appropriate alternative solutions.

Gerhard Plenert, Ph.D., CPIM

ACKNOWLEDGMENTS

In order to give credit where credit is due, I would need to create a long list of individuals, companies, universities, and countries that I have worked with. In my most recent academic past, I have had the pleasure of working with universities like the following:

- University of San Diego in their Supply Chain Management Institute
- Brigham Young University
- California State University, Chico
- Numerous international universities

Professionally, I have had the pleasure of working with organizations like the following:

- Wipro Consulting as a practice partner in supply chain management
- MainStream Management as a senior strategy and Lean consultant
- Infosys as a senior principal heading the Lean/Six Sigma/change management practice
- American Management Systems (AMS) as a senior principal in their Corporate Technology Group
- Precision printers as executive director of quality, engineering, R&D, customer service, production scheduling and planning, and facilities management

Other organizations that I have worked for include:

- Air Force and Department of Defense (DOD)
- The state of California
- The state of Texas
- United Nations
- And many more

I've lived in countries and worked in their factories, in the areas of Latin America, Asia, and Europe. I have co-authored articles and books and have worked with academics and professionals from as far away as Europe, Japan, and Australia. My broad exposure to a variety of manufacturing and service facilities all over the world has given me the background I needed to write this book on quality systems.

INTRODUCTION

Quality failures occur, not because the tools aren't available to solve the problems, but because the wrong tools are applied in the wrong ways.

GERHARD PLENERT

The author was invited to visit a high-tech company's factory. There had been a downturn in business, and the company decided to close the factory with the worst quality performance. The plant manager at the unfortunate facility had invited the author to express his opinion on why quality performance was bad.

"We have all the tools," explained the plant manager. "We have TQM (total quality management), SPC (statistical process control), TQC (total quality control), Six Sigma, work cells, etc. Why don't any of these systems work? Why are they failing us?"

"It's not the tools that fail," I explained. "It's the application of the tools that is failing you. Let me ask a few questions so that we can identify the problem. For example, do you have a bonus or incentive system for your shop floor employees?"

"Of course."

I continued. "How is it measured? What is the basis of payment?"

"Number of units produced."

"Number of quality units produced, or just number of units produced?"

"Strictly number of units produced," was the answer. "Quality is the responsibility of the quality department, not the line workers."

"So the employees are incentivized to push quantity through regardless of quality?"

"We have quality training programs," the plant manager informed me. "We have regular meetings and rallies where we focus on quality. Every employee knows about our quality goals."

I commented, "But they're not motivated to care about quality. It doesn't affect their bottom line."

The plant manager started to realize what was happening. "So you're telling me that unless I include a quality element to their bonus check, they're never going to care about quality? Are you telling me that all the quality training we've had has been a waste?"

"You need the quality training. That gives the employees the 'how' side of quality: How do they improve quality? But you're missing the 'why' side of the equation: Why should I care about quality? You're missing the motivator. I'm telling you that what you measure is what you motivate. I'm telling you that the most important performance measure is the paycheck, and anything that affects the paycheck will get the most attention. You can't delegate quality off to some 'department.' Quality is something that everyone in the company needs to be involved with. Quality is a team effort, not a department effort. You're putting up banners and posters on quality, but you're not putting your money where your mouth is. And until you measure quality on a personal level, you're not going to motivate quality at that same individual level."

The plant manager leaned back in his desk and said, "Oh!"

If quality is only given lip service and does not become a critical performance driver for your organization, then quality won't occur. Quality is not the responsibility of some obscure department or cost center; it is the responsibility of everyone: production worker, management, office clerks, quality department, and so on.

Often, the author will find himself in organizations where it takes two days to produce the part and ten days to process the order. That's not quality! Front office quality is just as important as production quality. Front office employees are the interface with the customer. A rude phone response can damage a company's reputation no matter how good the parts are. But the tools used to increase front office responsiveness and/ or throughput (which is often a measure of front office quality) may not necessarily be the same tools that you would use to improve production floor quality.

A different example from the author's experience may also be interesting to the reader. The author was brought into a different high-tech company. This time he was hired as director of quality. The company had a 14+ percent defect rate and the CEO of the company found that unacceptable. When the author came into the position, he discovered numerous problem areas. Similar to the last example, performance measures were a problem. But there were also problems in the front office, in order processing, in engineering, in shipping, and so on. All of these problem areas, combined together, resulted in a reputation of poor quality performance.

The author set out to define the specific quality failure areas. Everyone had opinions about where the quality failures occurred, and it was always "the other guy" that was causing the problems. The author took a hands-on approach and physically went to work in the production process, in the quality department, in the order processing department, and so on. He wanted to get on the ground and see what the employees had to say about the lack of quality. He wanted to get beyond the management rhetoric.

After a couple of weeks, it became pretty obvious where the problems were; they were everywhere. It was a combination of errors, process shortcomings, systems failures, and so on that resulted in the quality issues of this organization. And the author developed a plan of attack, which included a long list of improvement gaps:

- Identifying and implementing a production control system (MRP-based)
- Training for all the employees (especially management) in
  - Quality improvement processes and tools
  - JIT
  - Lean
- Changing the incentive system to incorporate a quality measure
- Engineering integration with the production floor (make engineers work on the parts they designed to see if they could design them better in terms of cost and ease of production)
- Reviewing all the front office processes
- Placing strategic focus on the R&D area
- Reviewing the supply chain integration

After the training was well under way (it's never finished), a quality week was initiated. The guiding principle for the week was "you can do anything to your work environment, change and rearrange any of the equipment, do whatever it takes, but you cannot produce any bad parts." It was basically a plant-wide Lean event. Originally, when the idea was introduced to management, they became concerned that they would lose valuable productivity and throughput. And, in fact, the first day was a productivity disaster. Employees were taking machines apart and rebuilding them, moving equipment around, and so on, and a considerable amount of throughput was lost. However, by the third day, the lost throughput was not only recovered—they were ahead of their normal three-day throughput. And they weren't producing bad parts.

The focus on quality, including quality training, and so on, continued on, and at the end of eight months the defect rate had been reduced to below 2 percent, and in following months it was brought to below 1 percent. In addition, cycle time had been reduced, inventory had been reduced by 40 percent, and capacity had been increased by 20 percent. The capacity improvement was the biggest surprise to management, but it should not have been a surprise when you realize that not wasting capacity in producing and fixing bad parts increases your capacity to produce good parts.

Was there one magical fix-all quality tool that was employed to solve all the problems in this factory? No! It was combination of several tools, appropriately applied, that eventually resulted in this success story. And there are numerous stories just like it.

The questions remain, "Which quality tool do you use where? How do you avoid the quality chaos that numerous companies are finding themselves tangled up with?" And that is the purpose of this book. It offers some guidelines for quality system and quality tool selection and implementation.

The purpose of this book is to provide private companies and government agencies with the tools to:

▲ Understand the power of quality tools
▲ Develop a systematic plan for the implementation of quality tools
▲ Understand the components of an integrated quality plan
▲ Understand the measures that define a World Class quality organization
▲ Develop a strategy and methodology to introduce quality principles

This book will enhance the reader's interest in developing and sustaining a competitive advantage by developing and sustaining leading-edge quality tools. The book will discuss how to integrate quality tools into an integrated, World Class environment. It will provide professional, objective, and valuable information to solve many of the major quality management challenges. It demonstrates solutions to these challenges by using stories and examples of how quality improvements have successfully made a difference in both the private and government sectors.

This book will focus on building great quality environments. It will offer a process for identifying and selecting appropriate quality tools, fill the gap that is being left by many of today's strategy books, and help organizations focus their quality plan by using the appropriate quality tools to fit a specific problem.

The structure of this book follows Figure I.1.

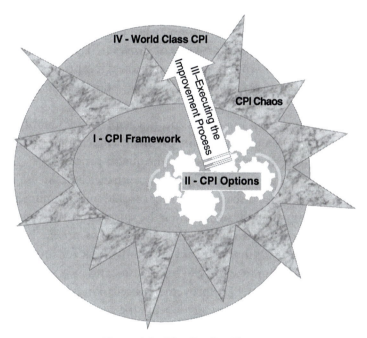

Figure I.1   The Quality Chaos.

Part I: CPI Framework   This first section discusses continuous process improvement (CPI) from a 30,000-foot level. What are the types of CPI systems? What constitutes a valid improvement project? What types of improvement should you expect?

Chapter 1 explores the purpose and value of this book and how its contents differ from traditional CPI books.

Chapter 2 discusses the characteristics of a CPI opportunity. What is a valid improvement opportunity? How should it be identified and treated?

Part II: CPI Options   This section delves into the details of how a strategic performance model needs to be properly selected and implemented in order to maximize the overall performance of the organization. This section also discusses the importance of selecting appropriate metrics and integrating them appropriately in order to achieve the desired organizational goals.

Chapter 3 explores the various quality options that exist and offers a benchmark for comparing them.

Chapter 4 delves into the characteristics of each of the quality options.

Chapter 5 maps the characteristics of each quality methodology against the types of quality problems that you may be experiencing.

Part III: Executing the Improvement Process   This section reviews the execution of the CPI improvements after the appropriate method has been selected. It discusses some of the logistics issues that are involved in CPI implementation.

Chapter 6 rounds off the discussion in Chapter 5 by identifying the appropriate tool to fit a specific quality problem. This chapter also emphasizes the critical role that metrics play in the quality process.

Chapter 7 describes the role of quality governance and the use of performance reviews. It also discusses how they cascade and communicate CPI improvement throughout the organization. It explains why this is important.

Part IV: World Class CPI   This section wraps up the book and stresses the need to continuously and constantly look for improvement opportunities if you aim to achieve World Class status.

Chapter 8 brings the CPI improvement discussion full circle and discusses the importance of continually moving the needle toward successful process optimization.

Chapter 9 summarizes the book and offers conclusions for moving forward.

PART I

CPI Framework

CHAPTER 1

Why Another Book on CPI?

Change sticks when it becomes "the way we do things around here," when it seeps into the bloodstream of the corporate body.

JOHN KOTTER

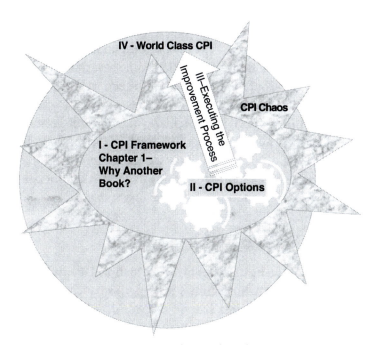

Figure 1.1 The Quality Chaos.

The author was driving through a Southeast Asian country with a friend to visit a new development area. There had been extensive construction activity over the last few years. A new town was built in a remote area of the region, and most of the roads and construction sites were too new to have been tracked by the various mapping systems that exist. After completing our inspection, we returned to the car and headed out of the area. A menagerie of roads and freeways had been constructed. The road signs were unreliable, so we started tracking our travel on the car's global positioning system (GPS). That system kept telling us to make turns that didn't exist. We were commanded to make turns into empty fields and to make U-turns on the freeway. So we turned on a portable handheld GPS. This second system kept giving us commands that contradicted the first GPS system, and both of these systems contradicted the few road signs that existed. We were left in the position of having the most sophisticated, leading-edge systems in the world, and getting completely lost. In the end we wiggled our way out of the situation, but our total travel time was increased by over one hour and we ended up in a completely different part of our destination city than we were originally shooting for. None of these systems gave us what we wanted, yet they were each considered to be "best in class."

This isn't too far off from the struggles we have with our current collection of quality control tools. Each is pointing in a different direction while simultaneously claiming that they know the best route for us to take in order to get us to our destination. And in the end, we entirely miss our desired goal. The end state isn't anything close to what we were looking for. Sadly, when we challenge the system, saying that it did not work for us, we are told that it didn't work because we didn't use it right. But, just as with the GPS system, it often isn't how we use the tool—the problem is that we're using the wrong tool. We're using a tool that cannot deliver the desired solution. We were doomed from the beginning.

Quality Methodology Performance

The author was visiting the manager of an international manufacturer who had become frustrated by his facility's poor performance. The company had plants at over 50 locations around the world. All these plant managers were trained in numerous quality management methodologies, and they had the posters on the walls and the books on their desks as tokens demonstrating that they were believers in quality. Yet their quality numbers were frustratingly poor. The cute slogans and quotes that they had posted everywhere didn't seem to help. And management expressed their frustration to the author. They blamed the lack of quality on poor worker commitment. But, in this case, it wasn't the quality tool that was the problem. It was indeed the misuse of the tool that created the problem. Signs and billboards do not make a quality system. It's the implementation of the system and the cultural transformation that will make the system a success. In future chapters we will discuss acceptance/change management tools. Without the cultural acceptance, we won't be able to make the type of quality transformations that we are targeting.

The West tends to look for the magic bullet. For example, when Toyota's high-quality standards were initially identified, everyone started scrambling to identify the "one magical piece" of the Toyota system that would explain everything. Then, once identified, the West was sure they could easily replicate it and thereby achieve and imitate Toyota's success. The author describes it as similar to "pill mentality." In the West, we constantly look for that magic pill that will supplement an unbalanced diet, or that will make us the perfect weight, or that will extend our life. There has to be a magic pill out there somewhere that will solve all and fix all. And that's the way we look at quality.

When the West analyzed Toyota, they started pulling out the buzzwords. First it was "quality circles." They believed that you just had to build quality circles and then you would have quality equal to the Japanese. When that didn't work, the next buzzword was statistical quality control (SQC). After that it was total quality control (TQC), and when that was identified as inadequate, it became total quality management (TQM). And it repeats on and on. Today's buzzwords are Six Sigma, which is an outgrowth of SQC, and Lean, which is an outgrowth and expansion of quality circles and TQM. And the question is: "Have we finally identified the magic pill for quality?" Unfortunately, the answer is "NO!"

If we go back to some of the original books published on the Toyota Production System (TPS) quality successes, we find stress placed on "cultural changes" rather than on "tools changes." The Western quality systems play lip service to "cultural transformations," but in reality they do very little to facilitate these transformations. Western adaptations of TPS jump almost immediately to tools, like Lean and Six Sigma, and ignore the foundation that makes these tools successful. Books like *Breakthrough Thinking* by Nadler and Hibino (Prima Lifestyles, 1998) or *Making Innovation Happen* by Plenert and Hibino (CRC Press, 1997) bring us back to TPS roots, teaching us how culture and the organizational focus are changed.

This book will address cultural/change management tools, often referred to as "acceptance tools." These tools are focused on the acceptance of change. This book will also focus on the "technical tools," which are the quality tools that we initially think of whenever we look for quality improvements. It's the magic pill of the West. However, both sets of tools are important and have appropriate applications.

Without cultural acceptance, these tools will not work. No matter how good the quality/change management tool is, if the organization's culture does not accept the tool as their own, it will fail. Similarly, no matter how bad a particular tool is, if the organization's culture has ownership in the tool, it will be a success. If they believe in it, it will work.

Today's quality world is plagued with a menagerie of methodologies, all claiming to be the "pill" that will solve all quality woes. This book will highlight many of these tools and will demonstrate how many of the tools were originally a piece of the TPS process. These tools were each independently thought to be the "pill" that cured all quality ills. And each was disappointingly replaced by the next generation of "pills." That doesn't make the tool bad. That just means that the tool wasn't the cure-all for all woes. The tool was designed for a specific purpose, and if we understand that purpose, we will know how to take optimal advantage of the tool. Each tool has its appropriate place in the schema of quality. Each tool also has areas where it is not effective. This book will attempt to highlight the plusses and minuses of each tool.

Defining CPI (Continuous Process Improvement)[1]

A business process is a well-defined sequence of steps, methods, and approaches related to transforming an input into a viable output. Business processes created around a continuous process improvement (CPI) methodology outline a specific direction incorporating relentless improvement

---

[1] This section was written with the help of Michael Thompson.

with a focus on business strategy. The results are well chronicled as business successes. This can be illustrated by the experiences of Toyota and General Electric.

The failure of business processes created around a CPI methodology also exist, perhaps best illustrated by the experience of 3M. The failures largely result from treating the methodology as a catch-all "silver bullet" to solve business problems without thinking about the specific strategic goal the methodology is supposed to address. In addition, the organizational culture required for the methodology to be successful is ignored. In order to fully appreciate the appropriate CPI methodology most appropriate for a specific organization, we need to compare the key features of each and compare and contrast the differences between the various popular CPI methodologies.

All methodologies examined here share several key characteristics and link specific business problems. Each has a goal with a set of metrics and/or guidelines that determine success. Each methodology came from a change in the market that required a new way to look at what was necessary to solve the business problems at hand. Each taps someone in the organization to be responsible for the improvement in order to see it through.

At this point we will look at some high-level CPI methodologies, which will be defined in more detail in future chapters. The first and most traditional of these methodologies is TQM. Many of the later methodologies are evolutions of this tool. Six Sigma, TPS, and Lean, for example, all evolved from TQM. TQM challenged the assertion in the early twentieth century that increased quality was the result of increased cost. Rather, increased quality through better processes led to increased productivity. Quality and cost savings could be achieved in tandem, but the processes needed to produce them needed to directly reflect the business metrics that are important to customers. TQM assigned responsibility to company management to rectify the quality problems of the organization. This came from a belief that most business problems flow from the unintended consequences of business managers' decisions. TQM emphasized that everyone bore responsibility for quality in some form. In later chapters we will review three variations of TQM that were developed.

The TPS came into practice in the aftermath of World War II. Japan lacked significant capital to redevelop its shattered industry. Utilizing TQM's focus on increased quality through better processes leading to increased productivity, Toyota developed the principle of "kaizen," or continuous incremental improvement. The intent was to decrease waste, or "muda," in the process. The best way to decrease waste according to the model is to put stress on the system to find its breaking points and challenge the bottlenecks. Toyota CEO Taichi Ohno directed the elimination of the "Seven Wastes" identified by Sheigo Shingo, detailed in Chapter 3. Work in progress inventory was minimized. The resulting transparency allowed management to pinpoint and address the bottlenecks accordingly using tools like theory of constraints, which led to a more efficient flow. At any stage in the process, workers were able to pull a readily available "andon" cord and stop the flow of goods in order to address the problems as they arose. Quality was more important than being on time in the production cycle. Timely addressing of production issues prevented producing defective products by avoiding the addition of value to an already flawed product.

Shingo also developed the Five Ss, also discussed in Chapter 3, as another tool to help employees fight waste. Machines were laid out in a circular pattern around the worker stations, which made sure that goods traveled less in the manufacturing center. Western factories of the time had long continuous production runs with little worker involvement. The circular work cells were seen as the key to efficiency. Smaller batch work was possible incorporating significant worker involvement and feedback. This became Toyota's key to efficiency, which was visible through improved cycle time and waste reduction.

In the 1980s, the Japanese threat to U.S. manufacturing required a reexamination of process improvement. U.S. firms realized that their long production runs resulted in significant errors and inefficiency when compared to the TPS. Motorola studied the TQM work of Deming and Toyota, and developed a method for improving quality through statistical process control. The intent was to measure and observe production and make in-process adjustments minimizing defects. The eventual goal became the elimination of waste to 3.4 defects per million opportunities. This would ensure that waste stayed within six standard deviations, or Six Sigmas, of the production mean. General Electric (GE) and its CEO Jack Welch championed the approach and attached a standardized process to it known as DMAIC, described later in this book. DMAIC stands for Define, Measure, Analyze, Improve, and Control. It instituted a team concept for continuous incremental improvement based on leaders trained in statistical process control. Trained individuals were awarded different belt levels based on proficiency. These teams of Six Sigma "Green Belts" led by Six Sigma "Black Belts" checked statistical controls in the GE processes and identified the root causes of problems using cause-and-effect diagrams, like fishbone charts. The reduction in variability from Six Sigma introduced increased quality into the production process.

Six Sigma appeared to be excessively dependent on statistical processes, which included significant capital expenditures for data collection. As a result, Lean became a major movement in the United States designed to cut costs. The elimination of waste and the decrease in capital held up in inventory became the major goals of the Lean movement. Process mapping became its most important tool. However, much of the methodology was a Westernized, less robust version of the original TPS. For Lean practitioners who were seeking to add statistical process control, Lean Six Sigma became increasingly important. It added some of the disciplines of Six Sigma, but with an emphasis on efficiency and elimination of waste.

Comparing CPI Methodologies

The CPI methodologies significantly overlap each other in terms of tools used, frameworks, and underlying ideology. Many are mutually reinforcing. For example, one of the Seven Wastes under the TPS is the production of defective products, and Six Sigma centers on preventing defects. A cynic might look at the similarities between DMAIC in Six Sigma and Plan-Do-Check-Act (PDCA) of Lean, or between the TPS and Lean, and say that we are just repackaging the same ideas. That would be an invalid statement. What each of these methodologies share is their development in response to a change in the market, such as increased competition from Japanese manufacturers or a paradigm shift in business operations. As a result, the CPI methodologies differ slightly in focus. This difference in focus is appropriate, since the market changes present different challenges to companies across industry and time. But the potential effects on a business are undeniable. General Electric states that the difference that Six Sigma brings to the firm is between $8 and $12 billion annually, while the TPS made Toyota the most successful car company in the world.

The development of the methodologies over time is significant, because as time progressed, tools and concepts grew more robust and formalized.

The requirements of an organization implementing Six Sigma are greater and more specific than the requirements of organizations implementing TQM. Complexity has increased over time, with the exception of the Lean methodology. A business trying to be more responsive to customers might adopt TQM, while a business trying to decrease waste might adopt Lean or the TPS. A business trying to increase quality might adopt Six Sigma.

For companies not ready to adopt a change in their organizational culture, many of these methodologies will be inappropriate. The level of employee involvement in the TPS, for example, giving any employee the ability to stop the production line with an "andon" cord is significant, and potentially more than many organizations can handle. Another example is that Six Sigma teams must make sure that they do not attempt to optimize their process by subordinating it to a constraint. The problem may be the constraint, and an organizational culture that does not allow the team to address the constraint will not realize the full benefits of Six Sigma.

Seeking to get control of manufacturing quality and costs and increase working capital, as well as improve quality standards, 3M brought in James McNerny from GE as CEO. Across-the-board adoption of Six Sigma principles did bring manufacturing quality and costs in line. It also destroyed the inventiveness of 3M's research and development organization and depleted the product pipeline. Stock values plummeted, and the new CEO reversed many of the Six Sigma initiatives. Given the difference in focus of the different methodologies, it is essential to note that the business must have a clear understanding of its requirements before implementing any methodology.

> There is only one boss. The customer. And they can fire everybody in the company from the chairman on down, simply by spending their money somewhere else.
>
> SAM WALTON

Summary

I was challenged to identify a good title for this book. Titles like "Quality Smorgasbord" and "Quality Conundrum" all seemed appropriate—because that is what we have: quality confusion. There are as many stories of quality system failures as there are of quality system successes. How is that possible? Aren't we taught that quality is always a good thing? Unfortunately, we have now learned that's not true. We need:

▲ The right quality methodology to fit our problem
▲ Implemented in the right (appropriate) way
▲ At the appropriate time
▲ With the right level of commitment (preferably from the top down)
▲ With the right facilitator (someone who knows how to appropriately implement the selected CPI)
▲ And with the right culture (change management oriented and committed) throughout the organization

That's a lot of "rights," and we have numerous examples of how missing any one of them has led to failure.

This book doesn't answer all those "rights," but it tries to give you a start by helping with the first one: what is the "right" system? The following chapters will guide you through the confusion of quality systems and head you down the road to answering all the other "rights."

> You don't think your way to change. You have to work your way to change.

CHAPTER 2

Identifying a CPI Opportunity

The definition of insanity is continuing to do the same things and expecting different results.

ALBERT EINSTEIN

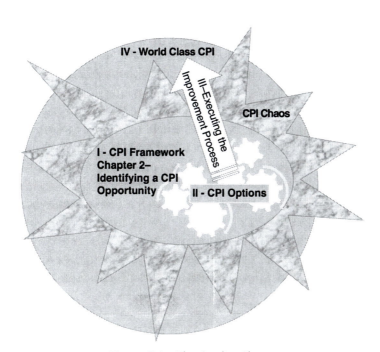

Figure 2.1 The Quality Chaos.

One of my favorite examples of change comes from one of the old episodes of the *Twilight Zone*. It's about a lady who was becoming frustrated with life. She was disappointed with life, was betrayed by her friends, lost her job, had an ever increasing number of bills, and nothing seemed to be going right. During a very frustrating and discouraging evening as she sat feeling sorry for herself there was a knock on the door. Opening the door she saw a stranger holding out a black box. In the center of the box was a button. The man handed the lady the box and told her that if she would press the button she would get a million dollars; however, "someone whom she did not know and who did not know her would die." With that the man departed.

Initially the woman thought this was the stupidest thing she had ever heard. She would never press this button causing someone to be killed, no matter how much money it got her. She put the box high on a kitchen shelf.

As time went by her discouragement started to gnaw at her. She reasoned that the person killed was probably really sick and was on the verge of dying anyway. Or maybe the person was a convicted murderer who deserved to die anyway. At one point she took the box down and nearly pressed the button in a moment of panic. But then, feeling a twang of guilt, she regained her composure and put the box back on the shelf.

As time went on she eventually lost her will to resist and in a moment of panic and extreme depression she rapidly took the box down and pushed the button. Almost immediately there was a knock on the door. It was the same man who gave her the box and he held out a check for one million dollars. Then he asked her for the box. The woman didn't realize that he would want the box back, but she got it for him and asked, "What are you going to do with the box?"

His response stunned her: "I'm going to give it to someone you do not know and who does not know you!"

Would that change your perspective? Wouldn't it have been nice to have an understanding of all the consequences of the changes you're going to make before you institute the change? Does it make you wish you had more data/information?

> Change and innovation are not stifled by the way things are, but by the way we incorrectly perceive things.

You've heard the saying, "The only thing certain in life is change." Even the rate at which changes are occurring is changing. Nevertheless, our nature tends to resist change. We like to be comfortable. We don't like confusion. We often want to resist it with all the effort we can muster.

There are two sources of changes:

1. The changes that come from us
2. The changes that happen to us

Treating the second source of change first, we need to be prepared for changes that will happen to us. Tsunamis in Indonesia and Japan destroyed countries and kill hundreds of thousands of people in the blink of an eye. We can't avoid them, but we can prepare for them. We need to watch for these changes and, whenever possible, manage them into opportunities. Change generates innovation; it triggers the development of leading-edge competitive strategies; it's moving forward.

The first type of change, change that comes from us, requires us to come up with the changes. For some managers, this type of change is difficult to manage because of the personal effort involved. Most individuals and organizations tend toward stability and bureaucracy, which suppresses change. For example, too many organizational structures are motivated by measurement systems that stifle and often punish change. We as individuals and corporations need continuous process improvement (CPI) systems that motivate the discovery of change. The search for change opportunities often requires both a structural reorganization as well as a mental/cultural reorganization of your enterprise.

This book focuses on quality methodologies that facilitate both types of change. It will discuss how your organization needs these quality systems in order to give structure and direction to your change process. Your organization needs planned, structured change, not just to maintain your status quo and survive in a competitive environment, but also to move out in front of the competition. The purpose of this book is to help you become a quality leader who can help your organization manage optimal change. Its goal is to give you the structure and tools that will motivate you to find and implement positive quality changes in a goal-oriented direction and to turn you into a leader who motivates continuous improvement. Finally, the purpose of this book is to help you become a world-class quality organization.

To get the most out of this book, carefully read the preface, introduction, and all the chapters. What you're doing is hunting for ideas. Mark the ideas that have the best fit, and, after reading the book, reread those ideas that impressed you the first time through. Then, once you've identified the quality methodology that will best serve the purposes of your organization, search for ways to implement these ideas into your organization. Reviewing this book regularly will help you reanalyze your quality change opportunities. Problems/opportunities change, and we need to look at the different opportunities with a new perspective, selecting the most appropriate tools for each situation.

> The single greatest power in the world today is the power to change . . . The most recklessly irresponsible thing we could do in the future would be to go on exactly as we have in the past ten or twenty years.
>
> KARL W. DEUTSCH
> PROFESSOR OF INTERNATIONAL PEACE, HARVARD

Models for CPI

There are numerous models for implementing change, from the slow and systematic process of total quality management (TQM), to the fast and radical tools like process reengineering (PR). There are CPI models that motivate change through their measurement process, and there are measurement systems that discourage change. Correctly implemented CPI models give us an entirely new focus on what change can do for us.

The author has spent many years dealing with technology transfer all over the world. He often asks, "How do you deal with change and innovation, such as the implementation of new technology in your environment?" Far too often the answer is, "Innovation is great and we like it, as long as it has already been tried somewhere else." That's similar to saying, "Change is good for everyone else, but not for me. It scares me." Another question often asked is, "How do you motivate innovation?" The attitudes of most people tend to be similar to:

> The innovator makes enemies of all those who prospered under the old order, and only lukewarm support is forthcoming from those who would prosper under the new.
>
> MACHIAVELLI
> THE PRINCE

Change is often forced upon us by internal problems and external competition, but problems and errors are also the seeds of opportunity and innovation. A world class manager (WCM) looks at these as opportunities rather than problems. Problems identify opportunities. As the Japanese manager mentioned in the introduction stressed, "No problem is problem!" This means that if you haven't identified problems, then you haven't identified opportunities for improvement. The WCM will resist "solving the problem" and will prefer to focus on the "opportunity for improvement." The "problem" can be solved by measuring and identifying its root cause. The "opportunity" to fix the root cause lets us build a better mousetrap, which will avoid the cause of the problem. For example, Post-it Notes started as a problem (attaching notes to a letter, book, or report without destroying it the way staplers or tape do). The problem became an opportunity (develop a new adhesive that allows for easy removal), and the opportunity has now been translated into a major product and market segment.

Small companies, and small countries, which are often the ones that are innovating, are upsetting large companies and countries, which tend to move slower. For example:

1. If you want process innovation, where do you go? For a long time Japan has demonstrated its international competitiveness through its ability to reduce production lead times and production costs. The Japanese were considered small at one time. Now they are considered to be the masters of process innovation.
2. If you want to introduce a new product into the market quickly, where do you go? Taiwan has become the time-to-market innovation leader.
3. Minimum cost and maximum flexibility in terms of steel production have been taken over completely by the small steel companies in the United States, which have, for a long time, been successful in beating out the big guys, both foreign and domestic.
4. And the examples go on endlessly.

One of my favorite quotes comes from the DreamWorks movie *Monsters vs. Aliens* where the U.S. presidential advisor says:

We need our top scientific minds on this. Get India on the phone!

Japan and Process/Quality Innovation

One of the best examples of turning a problem into an opportunity is seen in the Toyota Production System (TPS) and its tools, like the Just-in-Time production system. Just-in-Time (JIT) is a production planning philosophy developed in Japan that focuses on waste minimization through inventory reductions. But JIT didn't exist before being developed by Toyota. JIT wasn't copied. It was developed in response to a "problem." Here's a brief summary of the story.

It was post–World War II and Japan was trying to rebuild its industry. The Japanese tried copying Western (primarily United States) production methodologies, which were considered the best in the world, but they soon encountered the following four problems:

1. The Japanese lacked the industrial cash flow necessary to finance the large in-process inventory levels required by the U.S. batch-oriented production systems.
2. The Japanese lacked the physical land space to build large U.S.-style factories.
3. The Japanese didn't have the natural resource accessibility that the United States had, and materials would need to be purchased and imported.
4. Japan had an overabundant labor resource, which meant that labor efficiency systems weren't very valuable.
5. Japanese products were referred to as "Japanese Junk" because of their poor quality.

The Japanese turned these problems into opportunities. They recognized that their problem was a process problem, not a product problem. They proceeded to copy product technology from wherever possible, and they worked diligently to create quality and process technology focusing on materials efficiency rather than labor efficiency. The result was the TPS flow-through JIT production methodology, which included in-line quality inspections for which Toyota has now become famous. But Toyota will be the first to admit that it wasn't easy. Toyota officials scoff at U.S. attempts to copy JIT who pretend to be successful after two or three years of implementation. They will readily say it took them 30 years to develop JIT. But they got into this position one innovative change at a time.

The Japanese also scoff at the simple-mindedness of the way Westerners look at TPS. They readily share the current state of their production processes with the West because, in their mind, by the time Westerners successfully implement TPS or JIT, the Japanese will have innovated themselves so far ahead of their current state that the West will always be running behind them trying to catch up. TPS focuses on continuous process improvement. It's a never-ending battle to maintain their status as the leading-edge competitor.

The result was that the Japanese "Leaned" out their production processes. They built smaller factories (about one-third the size of their U.S. counterparts) in which the only materials on the production floor were the ones currently being worked on.[1] This kept inventory levels at a minimum. The focus was on materials (inventory) efficiency rather than labor efficiency.

Federal Express (FedEx), a Malcolm Baldridge Quality Award winner, is the brain-child of founder and CEO Frederick W. Smith, who single-handedly invented the air express industry. The focus of the FedEx quality program is to achieve 100 percent customer satisfaction through continuous improvement and change. The continuous improvement process involves the customer in the change process through a Survey-Feedback-Action (SFA) program.

FedEx has based its quality program on three precepts, the first being:

Customer satisfaction starts with employee satisfaction.

In order to make this precept effective, FedEx has implemented a program called the Guaranteed Fair Treatment Program (GFTP). The aim of the GFTP process is to maintain a truly fair working environment, one in which anyone who has a grievance or concern about his or her job, or who feels mistreated, can have these concerns addressed through the management chain, all the way to Fred Smith if necessary.

FedEx considers their employees their most important resource and wanted to provide a fair and equitable process for handling grievances.

---

[1] This statement is intentionally a little idealistic in order to dramatize the point. In reality, the Japanese work with single-digit batch sizes (one to nine units), whereas U.S. batch sizes can range in the hundreds of units. For each batch, only one item is worked on at a time; the rest of the batch is inventory. Therefore, a batch of 100 units creates a continuous, on-going inventory of 99 units. Unfortunately, the batch is often not being worked on and is just idle inventory. This batch size difference between the United States and Japan creates a tremendous difference in inventory levels.

The GFTP philosophy provides an atmosphere for employees to discuss their complaints with management without fear of retaliation. An employee is given seven days to submit a grievance, after which time management has ten days to respond. If the employee doesn't agree with the manager's decision, the employee has seven days to appeal the decision up the chain of the review process.

A key element of the GFTP program is that managers are evaluated from both directions: from the top down and from the bottom up. The manager's boss evaluates the manager's ability to implement change through innovation and improvements. The manager's subordinates evaluate the manager's responsiveness to the needs of the employees. A manager must receive favorable ratings from both directions in order to get promotions, raises, or bonuses.

From the FedEx quality program we see both an emphasis on change (SFA) and an emphasis on the management-employee relationship aspects of how change is implemented (GFTP). I've had FedEx managers tell me that FedEx is the most challenging, while at the same time the most rewarding, company they have worked for, because the FedEx program puts the manager into the challenging position of attempting to install change while at the same time not allowing the manager to be excessively forceful in implementing the change. The best managers would be those who implement change by giving the employees ownership in the change.

Goals

If you don't know where you're going, you'll probably get there.

Far too many people and companies go through life without targets. They let changes affect the road they take without ever identifying why they're traveling down the road to begin with. We need goals at many levels and at many time frames. We should have long-range (20+ years), mid-range (5+ years), and short-range goals (1 year). For example, what do you want to accomplish in the job function you are performing? Do you want to become world class? Far too often I have encountered people who, when asked, "How do you decide what areas of your job function you want to perform well in?" I get an answer like. "Whatever it takes to keep my job and get a raise." My reaction is, "I'm glad you don't work for me!"

After you have selected measurable goals, you now need to do two things:

- ▲ Develop an action plan that will move you toward the goals.
- ▲ Communicate the goals to all involved. For example, I have encountered numerous organizations where the corporate goals are pretty much kept secret among top management. Yet the employees are expected to achieve the goal. You can never overcommunicate your goals.

Your personal success needs to be defined. Most individuals need a more clearly defined game plan. There are several good books that help in the development of goals. Let me highly recommend the Covey and von Oech books if you haven't already read them.

Recently we have discovered a slight reversal of the goal-setting process in organizations. Previously the trend has been for the vision and mission of the organization to be defined by top management. However, we have seen a reversal of this process where the employees are defining the mission of the organization, and then a top management vision statement is developed from this employee-defined mission statement. For example, Tridon-Oakdale brought a team of managers together and had them establish the mission statement of the corporation. This gave the managers an ownership in the goals and an added commitment in achieving them.

We have a lot of people who are filled with good ideas, and we need to listen to them if we are to benefit from their wisdom. In addition, being world class means realizing that improvement only comes about if we open ourselves to changes in the form of new ideas. We need to learn new ideas in order to incorporate them into the things that we do. World class gives everyone in the organization opportunities and the appropriate motivation to learn and develop through education and training. We need to build a learning organization.

Chaos often breeds life when order breeds habit.

HENRY (BROOKS) ADAMS,
AMERICAN HISTORIAN

World class change management involves breaking out of the ritualistic, mundane things in life. Just because an employee is not an expert at something, doesn't mean he or she doesn't have worthwhile and valuable ideas. And the expression of all ideas, whether by the professional or the amateur, needs to be encouraged. The trick in managing ideas is to not allow egos to get wrapped up in the innovation process. Every idea has to be considered valuable, even if you really think it stinks, because you are a prejudiced observer, and you don't want to discourage the creative process. In addition, you need to be careful that the "professional" is not offended if the ideas of an amateur contradict his or her professional opinion.

Employees need to be involved in and to understand the change process in order to effectively initiate changes. This process is explained in many models similar to the following:

1. Identifying/recognizing the need or opportunity for change is the first step in making any change.
2. Define the problem or opportunity that needs to be addressed.
3. Identify the current company position relative to the problem—you have to know where you're at before you can determine where to go.
4. Identify alternative destinations.
5. Identify the desired destination.
6. Define a road map to get from where you are to where you want to be.
7. Unfreeze the organization and prepare it for change. This includes training and empowerment.
8. Implement change.
9. Stabilize the organization under the new order. This includes the establishment of a new feedback mechanism that will monitor the new status quo.

With an understanding of the change process, organizations and their employees are now ready for innovation and change creation.

This brings us to a critical element of world class employee innovation and creativity. We need to develop empowered work teams that work together as teams, not groups, and have the authority to implement their ideas. Creativity works best through the synergy of effectively developed and empowered teams. An example can be seen if we look at the Antilock Braking Systems (ABS) Division of General Motors in Dayton, Ohio (formerly Delco Products Company). They did what they were told would be impossible: they developed a world class empowerment program, called employee involvement (EI), starting with a traditional United Auto Workers (UAW) contract. A covenant was established between the union and ABS. The two felt this was the only way they would be in business in two years and that this was necessary if they were to stay even with the continuous improvement programs of their competitors.

ABS supervisors were given new responsibility based on communication and training. A system of trust was established with the workers. This trust is at the core of the EI program. Based on this trust, a set of guiding principles was established, which included:

1. We will establish and maintain innovative systems that can compete in a world class climate.
2. We will enact cultural change necessary to ensure profitability and job security at the Dayton plants.
3. We will run the business as a joint activity seeking contribution from and sharing benefits with all.
4. We will provide mechanisms and incentives which promote continual improvement in customer satisfaction.
5. We will approach this covenant as a living agreement, continually reviewing our progress and proactively adjusting to maximize our competitiveness.

ABS truly has a world class empowerment program that is worthy of study and emulation.

Successful Change Management

The business functions of an organization have, for a long time, focused on stability rather than on change. For example, IT, accounting, finance, personnel, the legal department, most upper management, and marketing would love nothing more than to have a steady, stable growth. Operations, traditionally, would love a perfectly balanced operation with just the right amount of inventory, just the right workforce, and no problems. However, one of the competitive lessons we have learned is that stability breeds failure. If we try to stay where we are, we'll get run over. Just ask the American passenger railroads.

But uncontrolled and undirected change can be as disastrous as no change. What we need is to be able to stay ahead of the change process. We need to change ourselves faster than external forces have a chance to change us. We need the change to be focused on a target. And we need to maintain our corporate integrity as we institute change.

To manage change, we need to incorporate change models into our business that facilitate the change process. Some of these change models, like total quality management (TQM) and process reengineering (PR), will be discussed in the next chapters. The problem with the change models is that they are often thought of as another fish story.

Companies (and change models) are like fish—after three days they stink.

Most change models contain some label of quality in them. Quality has become the flag behind which the battle for continuous change is most often fought. But "quality" doesn't fully define everything that is wanted by the change process. Terms like TQM and quality functional deployment (QFD) are change processes that look like they focus on quality; however, like all change models, they focus on positive, goal-directed changes in all the measurement areas, including quality, productivity, efficiency, financial improvements, and so on.

A model for change demonstrates the phases of growth in a change process. They are:

Phase I: Recognize the need for change. Invest in new technology or processes. Motivate innovation and experimentation. Encourage learning about new technologies just for the sake of learning.

Phase II: Learn how to adapt technology beyond the initial sought-after results. Keep the ideas flowing.

Phase III: The organization goes through structural changes as process changes occur.

Phase IV: The broad-based implementation of change occurs, affecting all aspects of the organization.[2]

The Japanese quality improvement methodology for the continuous change process is called kaizen. It suggests that every process can, and should be, continually evaluated and continually improved. The primary focus of the improvements is on waste elimination; for example:

Process time reductions
Reduce the amount of resources used

---

[2] Gibson, Cyrus F., and R. L. Nolan, "Managing the Four Stages of EDP Growth," *Harvard Business Review,* January-February 1974, p 76.

Improve product quality

And so on.

Kaizen problem solving involves:

1. Observing the situation
2. Defining the changes that need to take place
3. Making the changes happen

One example of the implementation of the kaizen continuous improvement process is at the Repair Division of the Marine Corps Logistics Base in Barstow, California. Utilizing the kaizen focus on continuous improvement, they ran a pilot project and they received results like the following:

63% reduction in final assembly lead time

50% reduction in work-in-process inventory

83% reduction in the distance material traveled

70% reduction in shop floor space requirements

The focus of any quality change model should be on continuous improvement in the broad sense, which includes both the Japanese incremental step perspective and the United States breakthrough business process improvement perspective. The need for change is rarely argued. What is different between the various change models is the speed of the change and the depth at which the change occurs. This is where the Japanese and the U.S. approaches to change methods bump heads. In a brief comparison:

United States	Japan
Fast change	Slow change
Fast return on investment	Long-term return on investments
Radical and dramatic change	Carefully planned out changes
Deep and extensive changes; feeling the need to redefine the whole process	Think the change through carefully. Plan before you implement.
On the hunt for the one big change that will fix all the problems	Small step changes
PR, which is characterized by rapid/radical changes and focuses on change implementation and high-tech solutions	TQM, which focuses on analysis and planning in the change process and technology-that-fits-the-situation solutions

(Continued)

United States	Japan
Slower to get around to making any change because the change process is viewed as being so extensive, dramatic, and upsetting. The result is that there is more resistance to any change process.	The change process is much less painful, because change involves small, undramatic steps. Therefore, there is much less resistance and, step-wise, small changes are continuously occurring.
Change ownership belongs to some change "hero" who quite often is the CEO	Change ownership is shared.

> Einstein was asked, if he had 60 minutes left in which to save the world, what would he do. His answer was that he would spend 55 minutes planning and 5 minutes implementing.

Some quality change methodologies have attempted, unsuccessfully, to combine the Japanese and U.S. approaches by suggesting the implementation of "radical changes without being radical." What they are hoping to do is implement big changes without upsetting the entire organization and developing enormous resistance to the change process. But no one has come up with a good way to accomplish this (probably because no one really understands it). So the conflict between the two change approaches remains. TQM and its companion, the Toyota Production System (TPS), continue to be viewed as "too slow" by the United States, and PR continues to be viewed as "too destructive" by the Japanese.

Summary

In conclusion, before CPI and change management in a quality transformation can achieve significant improvements, there must be an acknowledgement and commitment from the organization's leadership. There needs to be leadership acceptance throughout the organization that the current process that they are now running either is not or in the future will not meet the needs of customers and, therefore, will not ensure the continued survival of the organization. Once this acknowledgement is made we can get the organization committed to change. And with that commitment to change the quality methodologies evaluated in the next few chapters will facilitate the desired change processes.

PART II

CPI Options

CHAPTER 3

What CPI Options and Benchmarks Are Available?

I'm always doing what I can't do yet in order to learn how to do it!

VINCENT VAN GOGH

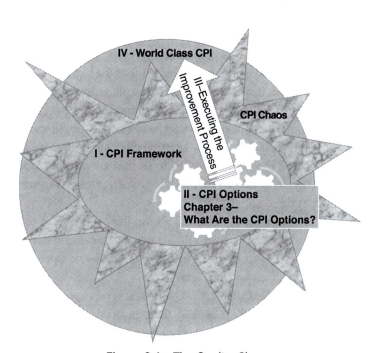

Figure 3.1 The Quality Chaos.

This chapter starts us on a journey of never-ending alternatives. Not only are there numerous continuous process improvement (CPI) alternatives, each claiming to have the magical solution to all our problems. But many of these CPI methodologies are interrelated subsets of each other. For someone new to the CPI world, this can become extremely frustrating and confusing. Even for those of us who have worked many years in this area, it gets unwieldy very quickly. And, like the Cheshire Cat's recommendation, we quickly feel like it really doesn't matter what road we pursue—all roads reportedly lead to quality. Unfortunately, it really does matter which road we follow, because the roads eventually lead us to different quality destinations, and they are not always the desirable destinations.

This chapter develops an arbitrary list of current CPI fads. It does not attempt to be all-inclusive and incorporate old fads that have fallen by the wayside. Rather, this list is focused on tools that are currently being used. It is structured by group, since many of the tools are not really unique tools, but are often subsets or subelements of other tools. In this chapter each of the tools will be defined and briefly explained. This book does not attempt to give you in-depth details about each of the tools. Rather, this book is designed to point you to the correct tool, and then the reader will need to explore the tool in more depth by additional reading or training.

So let's press forward by listing and defining the quality tools that are popular in today's CPI environment. Table 3.1 lists the high-level tools that we will be discussing in this book. But a little explanation is required first. For example, item 1 includes all the Toyota Production System (TPS) tools, which includes numerous buzzwords that have become popular in today's CPI environment, like TQM, TQC, Lean, Six Sigma, QFD, and concept management. Items 2 through 7 are CPI tools that are not offshoots from TPS. They are entirely different methodologies. Each major segment will be listed as a separate section of this book.

Diving deeper in the TPS, Table 3.2 shows the collection of tools that are outgrowths and offshoots from TPS. As you can see, the list of offshoot tools is quite extensive, and in order for this book to be complete, each of these will need to be discussed and analyzed.

In Table 3.2 we see that Item 1.1, TQM, has several subdefinitions. This occurs because several individuals had their own interpretation of what TQM is and how it should be used. And each of these techniques is actively utilized. Item 1.3, Lean, lists a large number of tools that may be used as

Table 3.1   Quality Methodology

1	TPS—Toyota Production Systems
1.1	TQM—Total Quality Management
1.2	TQC—Total Quality Control
1.3	Lean
1.4	Six Sigma
1.5	QFD—Quality Functional Deployment
1.6	Concept Management
2	Breakthrough Thinking
3	TOC—Theory of Constraints, Bottleneck Analysis, Constraint Analysis
4	Process Reengineering
5	ISO Standards
6	Strategic Mapping
6.1	Hoshin Planning
7	Simulation/Modeling

Table 3.2   Quality Methodology

1	TPS—Toyota Production Systems
1.1	TQM—Total Quality Management
1.1.1	TQM—Deming Version
1.1.2	TQM—Crosby Version
1.1.3	TQM—Juran Version
1.2	TQC—Total Quality Control
1.3	Lean
1.3.1	Process, Project, or Event Charter
1.3.2	Scan
1.3.3	Kaizen Events, Rapid Improvement Events (RIE)
1.3.3.1	PDCA—Plan Do Check Act
1.3.3.2	A3 Reporting
1.3.3.2.01	Root Cause Analysis
1.3.3.3	Acceptance Change Management Tools
1.3.3.3.01	Change Acceleration Process (CAP) Model
1.3.3.3.02	Kotter Change Model

(Continued)

Table 3.2 Quality Methodology (*Continued*)

1.3.3.3.03	Myers Briggs
1.3.3.3.04	JoHari Window
1.3.3.3.05	Ladder of Inference
1.3.3.3.06	PAPT
1.3.3.3.07	Situational Leadership
1.3.3.4	Technical Tools
1.3.3.4.01	Value Stream Mapping
1.3.3.4.02	Seven Wastes
1.3.3.4.03	System Flow Chart
1.3.3.4.04	Swimlane Chart
1.3.3.4.05	Spaghetti Chart
1.3.3.4.06	5S
1.3.3.4.07	Cellular Work Design
1.3.3.4.08	JIT—Just-in-Time
1.3.3.4.09	Agile
1.3.3.4.10	Poka-Yoke
1.3.3.4.11	PQ Analysis
1.3.3.4.12	TPM—Total Product (Productive, Preventative, or Production) Maintenance
1.3.3.4.13	Visual Workplace
1.3.3.4.14	DFM—Design for Manufacturability
1.3.3.4.15	SMED—Single Minute Exchange of Die, Quick Changeover
1.3.3.4.16	Kanban
1.3.3.4.17	Jidoka
1.3.3.4.18	Standard Work
1.3.3.4.19	Brainstorming
1.3.3.4.20	Fishbone Charting
1.3.3.4.21	Eight-Step Problem Solving
1.3.3.4.22	SWOT—Strengths, Weaknesses, Opportunities, Threats
1.3.3.4.23	VOC—Voice of the Customer
1.3.3.4.24	Gemba Walk—Go and See
1.3.3.4.25	Gap Analysis
1.3.3.4.26	Five Whys
1.3.3.4.27	SIPOC or COPIS

part of a Lean process, but many may also be used as part of a Six Sigma (Item 1.4) process. In order to avoid duplication, these tools are only listed once under Lean and are not repeated in the Six Sigma section.

1   TPS—Toyota Production System

The Toyota Production System is not a tool; it's a philosophy focused on continuous improvement. There isn't a magic button that you press, and now you have incorporated the TPS into your current quality environment. TPS is not fixed in time. It's constantly changing and evolving. What we knew as the TPS 20 years ago is not the same as the TPS of today. However, the West (North America, Europe, Australia, etc.) have constantly been looking for the magic bullet that will allow them to become as good as the TPS in terms of quality.

I had a conversation with a Toyota manager where I asked him, "Why are you willing to share all your quality/CPI secrets with the West?" His answer was honest, and interesting. He said (I fixed some of the English), "As long as the West is trying to learn our technique, that means they're behind us. The only way they're going to beat us is by jumping out ahead of us with something new and innovative. But they haven't figured that out yet. For now they're perfectly happy copying us, which means they'll always be behind us and we'll always be ahead." He's saying that TPS is a constantly moving target, always improving, always innovating, and always changing. And a copier will never be a leader.

A different, but also interesting, exchange with a Toyota manager occurred during a Toyota quality meeting in a U.S. facility. The visiting Toyota manager asked the plant manager what problems he was having in the plant. The manager answered that he wasn't having any problems; everything was running smoothly. The visitor responded with, "No problem is big problem." After some additional explanation we realized that he was telling the plant manager that if the manager wasn't looking for ways to improve the facility, than he wasn't value added; he was a waste and should be eliminated. The manager's role was to identify problems (opportunities/shortcomings) and focus on identifying solutions. The plant should never be "running smoothly." That wasn't CPI. That wasn't TPS.

So now I'm left with trying to define TPS and that's challenging. TPS is a culture, it's an attitude, it's a philosophy, it's a way of thinking, and it's a collection of tools that were developed to support the culture and philosophy. The West focuses on the tools, while the East focuses on the culture and philosophy. That's the difference. The tools are stagnant, while the cultural perspective adapts. The basic principles are as follows:

▲ The only value-adds in any system or process are those things that directly thrill the customer; everything else is a waste.

▲ Every employee's primary role in a company is to continuously look for opportunities to increase value and eliminate waste.

▲ The company's reason for existing is to give their employees and their families a better life. This is accomplished by controlling market share and destroying the competition.

▲ Everything is a system and a process, and as such can be optimized and improved.

▲ A successful company must have a basic foundation of trust, both within the organization with its employees, and without in its relationship with customers and suppliers. Therefore, control systems are evidence of a lack of trust and are a waste.

▲ Measurement tools and metrics are used for motivation, not for data collection. A measurement that does not promote a response focused on increasing value-added content and eliminating waste should be eliminated.

▲ Optimize the throughput and value-added content of your critical resource (which is usually materials, machinery, or the logistics network), not your employees.

⚖ Anything short of perfect quality wastes resources and capacity, and is itself a waste.

⚖ And so on—you get the idea. This type of thinking is still foreign to the Western manager.

This collection of statements is about as close a definition of TPS as we can get. This is the philosophy upon which is built a foundation of tools that are used to execute these philosophies. And this book will focus on a few of the philosophical elements. However, since this book will primarily be read by Western managers, this book will mostly focus on the tools.

1.1   TQM—Total Quality Management

Total quality management (TQM) wasn't the first tool extracted from TPS. But earlier tools, like quality circles, have been dropped from the vocabulary and are no longer in common use. Therefore, these older systems won't be considered in this book. TQM is still alive and well in various parts of the world. In the United States, TQM has been given a bad rap, primarily because of a misuse of the tool, and it has been repackaged and renamed. But the foundational principles of TQM are still valuable and effective, as evidenced by the countries that still actively use it.

In this section we will discuss TQM in general. Then we will focus on some of the variations of TQM in how it is applied. In Table 3.3 we see the aspects of TQM that we will be discussing.

TQM is one of the most popular CPI models internationally. It is extremely popular in Asia and Europe, primarily because it focuses more on analysis than on speed. It is the reverse for process reengineering (PR). Numerous companies throughout the world have focused on the TQM change model. Perhaps the first, and most impressive, U.S.-based global

Table 3.3   Quality Methodology

1	TPS—Toyota Production Systems
1.1	TQM—Total Quality Management
1.1.1	TQM—Deming Version
1.1.2	TQM—Crosby Version
1.1.3	TQM—Juran Version

company is Motorola, which has several continuous change programs throughout their plants all over the world. They are using quality councils and empowered teaming, both basic tools of the TQM process, in order to motivate their employees toward change.

Following the Motorola example, more by force than by desire, are many of Motorola's vendors. One of these vendors, and probably the most successful implementer of TQM, is AT&T, which is another global company with a focus on global efficiency and change. AT&T has systematized their change process to the point where they have corporate-wide documentation of how the system works. The result of the AT&T implementation has been that they are only the second U.S. company to win the Deming Award (the Japanese award for quality change). In addition, they have won three Baldrige Awards (the U.S. award for quality improvement) and the Shingo Award (the U.S. award for excellence in manufacturing).

TQM is not just a tool—it is a philosophy about how businesses should be run. The philosophy of TQM is filled with ideas and attitudes. Basic to this philosophy is the idea that the only thing certain in life is change. And we can either wait for the change to happen to us, or we can become an instrumental leader in the changes that will occur anyway. Competitive edge is rampant with changes. Product life cycles are shortening, and the lead time to market for new products is becoming increasingly shorter. We may have the best ideas and products in the world, but unless we can get them to the market, in a way that will appeal to the customers, quicker than our competition, what's the use? The following is a list of several of the key elements of TQM philosophy:

Desire change and seek it out
Think culture—move from copying to innovating
Do the right things before you do things right
Use a top-to-bottom corporate strategy
Have a clear definition and implementation of quality
Educate, train, and cross-train
Integration and coordination are key
Have an international viewpoint

Change is the universal constant. In TQM, the philosophy behind change is one that suggests that we become excited about change. We look for opportunity to change, especially because change should mean that we are becoming better. To be a TQM organization is to become an organization that wants to be the best, and realizes there is always room for improvement.

The idea behind a "think culture" is that we become a company that moves away from copying to innovating. This type of organization realizes that the best you can ever do by copying someone else is to get caught up with them. A think culture is where we know that in order to be the best we need to innovate. We need to create and then implement our new creations. We cannot be afraid of taking the risks that innovations will require of us. The concept of a think culture also suggests that we need to break free of the fads. ERP (enterprise resources planning), JIT (just in time), OPT (optimized production technology), TOC (theory of constraints), CIM (computer integrated manufacturing), BAM (bottleneck allocation methodology), and so many more, are all fads that have been sold to us as world cures. A think culture looks at each fad as having potential benefits, but realized that nothing fits everyone. These fads have to be looked at in light of the goals that you are trying to achieve. They would be implemented as stepping stones toward improvement, not as stand-alone, complete success packages. Often, they will be found to be lacking in the benefits that they were supposed to have.

The idea of a think culture also follows closely with the next concept, which I have labeled "Do the right things before you do things right." This means that you need to consider the value of what you are doing (or planning to do) before you do it. There is no value in doing a job perfectly if the job adds no value to what you are trying to achieve: your corporate goals. All activities in an organization need to be looked at in light of their added value to the organization. This includes rethinking what you are already doing, as well as taking a close analytical look at innovations.

TQM supports a "top-to-bottom corporate strategy," which means that the direction for the corporation comes from the top. It means that without top management commitment to change, as well as a commitment to the other philosophies supported by TQM, TQM is doomed to failure. This top-to-bottom strategy also supports integration between the various levels of the organization.

TQM philosophy supports a clear definition and implementation of quality. There are nearly as many definitions of quality as there are companies. A TQM company needs to define quality for itself, whether that definition is to meet engineering standards, as it is in most U.S. companies, or whether it is the leading-edge definition of making a product so exciting that the customer wouldn't think of buying from anybody else but you. With a clear definition of quality the company can start to focus on a target for change. It is difficult to focus on implementing changes that will improve quality if no one agrees on the definition of quality.

Another TQM philosophy is the continual need for education, training, and cross-training. Change and innovation look for the implementation of new ideas. Searching out these ideas and validating their usefulness require an open-mindedness and a constant search for new technologies. This requires education. Then, when ideas are selected for implementation, the implementers and users of the new technology need to be trained in their application. In addition, in order to be an effective, integrated organization, cross-training (understanding the jobs and functions of coworkers) allows employees to help other employees, as well as to see suggested changes "through their eyes."

TQM philosophy states that without integration and coordination of functions, you are ineffective in the exchange and implementation of new ideas. This integration must be both horizontal (between functional areas) and vertical (top to bottom) in the organization. A free exchange of ideas between functional areas without constantly being concerned about the chain of command is critical.

The last philosophical point listed is the need to realize that all companies are affected by international transactions, whether through the vendor, the customer, or directly. An international viewpoint when it comes to the development of new ideas can mean the difference between a competitive and noncompetitive edge because it opens up a whole world of markets and vendors.

The success stories for TQM can be found in settings all over the world. This success is measured in terms of how well change is implemented. This change can take the form of a new technology or the correction and improvement of an old technology. A successful TQM project results in employees able to work more effectively together.

Now we will discuss what is needed to take the TQM philosophy and make it operational. Several points need to be discussed. The first, goal focusing, emphasizes that the majority of the employees of an organization have no idea what the goals of an organization are. They will offer some guess, like profits or sales, but they are guessing. Obviously, if the employees don't know what the goal of the organization is, then how should they know if what they are doing helps to achieve this goal? If we are considering the implementation of a TQM program, then all of the TQM teams need to evaluate the projects being considered for implementation in light of this goal. We need to determine which project contributes the most toward achieving the goal.

TQM implementations start with a coordinating team, often referred to as a quality council. This is a team composed of high-level corporate leaders from all the functional areas. This team is appointed by the CEO and operates under his or her direction. The CEO takes an active part in directing the activities of the team. This quality council is then responsible for organizing and measuring the performance of the other TQM teams within the organization. It oversees the installation, training, performance, and measurement of the other teams. This team focuses specifically on the corporate goal/vision and definition of quality.

The quality council will organize three different types of teams, referred to as the cross-functional three "P" teams. The three P teams are process, product, and project teams. The process teams are on-going, continuous improvement teams set up at different levels of the organization. They look for improvements in the organization's functioning processes. These teams should be composed of both "insiders" and "outsiders." The insiders know and understand existing functions and operations. The outsiders challenge the status quo.

The second of the three P Teams are the product teams. These teams are cross-functional but focus on a specific product or product line. They are customer and vendor interface teams that are specifically oriented toward the development of new products and the improvement of existing products. Their life span is the same as that of the product they represent.

The third of the three P teams are the project teams. These teams are limited-life teams set up to specifically focus on a specific project, like the construction of a new plant or a computer installation. These teams may be the result of a specific process or product that is being targeted, or they may be set up to research something that the general management team is interested in developing or improving.

At this point we should know where we are going as an organization (focused goals), how we are performing toward this goal (measurement and motivation systems), and who is going to help us get there (TQM teams).

Next we need to discuss the processes that a typical TQM team will go through. These are called the TQM project implementation steps:

▲ Identify problems (opportunities)
▲ Prioritize these problems
▲ Select the biggest bang-for-the-buck project
▲ Develop an implementation plan
▲ Use operations research and management information system (MIS) tools where appropriate
▲ Develop guideposts and an appropriate measurement system
▲ Training
▲ Implementation
▲ Feedback—monitoring—control—change
▲ After successful project implementation and on-going status reports, repeat the cycle

The first job of the team is to identify their function and charter. If you are on one of the three P teams, your team's charter is laid out for you by the quality council. If you are the quality council, this charter is laid out for you by the CEO and is aimed at the focused goals of the organization. Understanding your charter, the team then searches for and identifies problems that exist and that prevent the organization from achieving this charter. The word "problems" has a negative connotation. It would be better to say that we search for "opportunities for improvements." We are not just trying to correct negative effects—we are looking for techniques or tools that will allow us to become better and possibly even the best.

Next we take these problems (opportunities) and prioritize them based on their effect on the charter of the team (which should be focused on the goals of the organization). We do a type of ABC analysis (80-20 Rule or Pareto principle) to determine which change would have the greatest effect. Then we select the biggest bang-for-the-buck project and develop an implementation plan for it. This implementation plan needs to contain guideposts that are based on an appropriate measurement system that points the team toward achieving its charter. The book *Breakthrough Thinking* by Gerald Nadler and Shozo Hibino (Prima Lifestyles, 1998) does an excellent job of discussing opportunity identification techniques.

Training of the implementers and users is critical, or else the planned project is doomed to failure. This training makes future users comfortable with the changes. It also offers a bit of ownership, since the users will now feel comfortable with the changes.

The next step is implementation. The implementation should be trivial, if all the planning and training steps are performed carefully. Part of the implementation is the installation of feedback, monitoring, and control mechanisms. Careful monitoring allows for corrective changes to occur whenever necessary.

After successful project implementation and seeing that the on-going status of the project is functioning correctly, the team repeats the implementation cycle, looking for more new opportunities for change. If this process is performed correctly, the list of change opportunities should become longer with each iterative cycle. This means that your team is now open for newer and broader opportunities for change.

Training programs need to exist before and after project selection. In the before case, the TQM team needs to understand what tools are available to them. This training would involve an understanding of tools and techniques.

There are three TQM "gurus," each with a strong opinion on what's important in order to achieve world class quality. Unfortunately, they also disagree violently with each other. Each has developed a strong following and has gained respectability because of their success. We will briefly explore each of their TQM methodologies.

1.1.1   TQM—Deming Version

Dr. W. Edwards Deming went to Japan to do some statistical quality training long before the West came up with the name TQM. Japan had developed a reputation for "Japanese junk," a reference that has long since been forgotten. In order to change this label, Japan embarked on a campaign to become a quality leader. Deming was a part of that campaign and had such a dramatic impact on the quality processes in Japan that the Japanese award for quality is called the "Deming Prize."

Deming's approach followed several key principals, such as:

- ▲ Statistics is the key to eliminating process variation.
- ▲ Management needs to own and feel responsible for quality failures.
- ▲ Continuous improvement initiatives are driven from the top.
- ▲ Variations are the root of all failures and need to be eliminated.

The Deming model of TQM focuses on three types of quality:

1. Quality of Design   This requires an understanding about customer expectations. How does the customer define quality?
2. Quality of Conformance   This requires an understanding of how the firm and its suppliers can meet and exceed customer expectations. How can we surpass the design specifications?
3. Quality of Performance   This looks at how the product performs. How can we redesign the product to perform even better?

Deming built his TQM plan around what is now referred to as "Deming's 14 Points." According to Deming, by following these 14 points an organization can work its way to becoming a world class quality leader. The 14 points are:

1. **Create consistency of purpose toward improvement of services.** What Deming is referring to here is a mission statement, which is a living document clearly defining the direction of the organization, and which is understood and has "buy-in" by everyone in the organization. The mission statement should address:
   a. People
   b. Customers
   c. Suppliers
   d. Community
   e. Vendors
2. **Adopt a new philosophy.** This requires an acknowledgment that the current state of performance is no longer acceptable. Defects are costing us money, and that's not acceptable. Quality must be built into the product, not inspected out later. Everyone in the organization participates in quality. The only acceptable definition of quality is the customer's definition.
3. **Cease dependence on mass inspection.** Control the process though statistics, and fix problems as they arise. Don't wait for inspection—by then it's too late. Inspections don't fix quality problems because they are too late to have an impact.
4. **End the practice of awarding business on the basis of the price tag.** Business should be awarded by using a combination of quality and price as measures of performance. Technical specifications should also include performance specifications. Procurement needs to understand

statistical quality control (SQC) tools and should actively use them for their inspections of received materials.

5. **Find problems.** Management's role is to identify problems and move their organization toward finding solutions to those problems. Deming has identified two types of process variations, both of which are eliminated using statistical tools:

    a. Special variations   Variations that are under the control of the operator and where the operator needs all the relevant information so that he can take corrective action.

    b. Common variations   Variations that are "common" to the system itself and where the operator has no control. Deming claims that 94 percent of all variation is caused by common variations, which require system revamps in order to solve. Common variations include things like:

        (1) Hasty, quick-fix product or systems designs
        (2) Inadequate inspection of incoming materials
        (3) Failure to understand the capabilities of a process
        (4) Poor lighting, dirty work environment
        (5) Failure to supply adequate statistical signals for the workers

Deming stresses that "a process is in control only when it can be controlled by the worker." It is management's responsibility to eliminate all common variations through a major systems overhaul so that control can be turned over to the operator. Tools that are used to improve or eliminate common variations include:

1. Control charts
2. Scatter diagrams
3. Histograms
4. Fishbone charts
5. Flow charts
6. Brainstorming

Once a stability has been achieved, workers become accountable for special variations that occur in their processes.

6. **Institute modern methods of training.** By "modern methods," Deming is referring to principles like:

    a. Align training with the objectives of the firm
    b. Identify goals that will be met through the training

    c. Train everyone on statistical thinking

    d. Focus on team building

    e. Develop training programs only after analyzing what needs to be taught

7. **Institute modern methods of supervision.** By "modern methods of supervision" Deming is referring to not penalizing employees for "systems" problems or common variations. Management is not the "policeman" of the employees. There should be a relationship based on trust.

8. **Drive out fear so that everyone may work effectively.** This is about changing the work climate away from one of fear to one of trust, where the employees feel they have the power to change different aspects of their lives.

9. **Break down barriers between departments: everyone must work as a team.** This point targets the elimination of organization silos where each organization sees itself as an independent entity from the other organizations. Barriers of administrative levels, communication channels, and quotas that create competition create contention between the organizational elements and need to be eliminated.

10. **Eliminate numerical goals, posters, and slogans that seek new levels of productivity without improving the methods.** Deming does not believe in slogans like:

    a. Zero Defects

    b. Quality Is Job One

    c. Increase Profits or Sales

    d. Be Careful

    He thinks that these are the responsibility of management and that tormenting the employees with these statements is meaningless. The goal of management is to create a stable process by eliminating common variations. Until that happens, the employees really have no control.

11. **Eliminate work standards that prescribe numeric quotas.** Deming feels that quotas and MBO (management by objectives) are not conducive to improvements. He feels that these metrics are arbitrary and are an attempt to legitimize and break down management's "grand plan." The problem is that employees are not given new tools to accomplish these quotas; they are expected to continue the process as usual, yet get

better. Arbitrary quotas drive employees to focus on volume, rather than on quality.

12. **Remove barriers that rob employees of pride in workmanship.** Since management does not work with employees to improve their processes, the employees feel that their input has no value. Arbitrary barriers have been established by management that prevent "pride in workmanship," like:

   a. Lack of management understanding concerning common and special variations

   b. Performance appraisals that destroy a team perspective

   c. Production reports that focus on historical volumes without acknowledging variations

   d. Failure to understand visible and hidden costs in cost of quality

   Deming believes that performance appraisals should be eliminated. Promotion needs to be team-based and focused on how an employee performs as a team member, rather than as an individual.

13. **Institute a vigorous program of education and training.** Deming believes in statistically based training. Additional training should be supplied in communications. In general, he feels that a broadly educated workforce is the most effective.

14. **Create a program that will push the prior 13 points each day for never-ending improvement.** Deming believes in creating a cycle of management-driven continuous improvement. The leader of this process should be a statistical expert helping management to focus on statistical thinking.

1.1.2   TQM—Crosby Version

Crosby's approach to quality improvement is less statistical and more cultural. He focuses on the cultural aspects of the management, employee, customer, and supplier relationships. For him, a low-quality firm is one where

▲ Employees complain a lot

▲ Customer complaints are on the rise

▲ Management direction is vague with a minimal focus on quality

▲ Direction focuses on schedule and cost rather than on quality

▲ Management denies its role in quality problems and does not make meaningful changes

Crosby's claim-to-fame slogan is "Zero Defects (ZD)," which, as we saw earlier, contradicted Deming's point 10. Crosby's focus is that anything less than ZD is not acceptable. The entire organization needs to be committed to this goal. But employees need to be motivated. And poor management practices that cause employees to be demotivated include

▲ Performance reviews that are routine and not meaningful or helpful
▲ Excessive expense accounts
▲ The attitude of "the boss rules the roost"

Crosby believes that irritated employees do not produce quality. They tend to focus on "self-protection." Motivated, appreciated employees do produce quality! They tend to not be as defensive.

Crosby's definition of quality is "conformance to requirements." He recommends setting the highest possible standards, and then driving toward conformance to these requirements. He recommends using SQC to identify critical process variables, which need to be measured so the process can be adjusted as needed. His focus is on employee-level process control, which is challenging for Western organizations to accept.

Another Crosby buzzword is "DRIFT," or Do It Right the First Time. This supports the ZD concept of only doing high-quality work. This is the only acceptable performance standard. No level of defects is acceptable. All mistakes are the result of a lack of knowledge or a lack of attention to the process.

1.1.3 TQM—Juran Version

Dr. Joseph Juran feels that the issue with quality is "How do we do it?" He feels that the failure is with management. It's management's responsibility to learn how to manage for quality. He believes that quality needs to be managed as if it was equal in importance to any major financial problem. It should be evaluated using a similar set of three steps, which are

1. Financial planning: Set financial goals and a plan of action
2. Financial controls: Evaluate on-going performance and compare it with the actual goals
3. Financial improvement: This improvement is continuous and should be better than in the past

Organizations treat quality as a "start-up" program where it is just a quick-hit process. Quality needs to be treated similar to the three-step financial process, with a long-term, continuous improvement perspective, not just as a "monthly meeting."

For Juran, quality is defined as "meeting customer expectations." This is translated into the three-step process, which is as follows:

1. **Quality planning:** Set quality goals and develop a roadmap for achieving these goals. A poorly planned set of quality goals and roadmaps is the foundation of poor quality problems. Quality planning requires the following:
   a. Identify the customers
   b. Determine customer needs
   c. Define product features to meet these customer needs
   d. Establish a set of product goals based on the customer requirements
   e. Define process improvements that will meet the product goals
   f. Establish a feedback mechanism for future action
2. **Quality control:** Fundamental to quality control is self-control, which means that there are systems tools in place that will let people know what their quality goals are and how well they are performing to those goals. It also requires that employees have the ability to adjust their processes so they can come closer to achieving their goals.
3. **Quality improvement:** Top management is involved in organizing project teams, and then in actively participating as members of the quality improvement teams. Management needs to help employees resist the feeling that "this is just the latest fad." The improvement process focuses on projects, and is taken one project at a time. The improvement process is not treated as an enterprise-wide breakthrough; rather, the focus is on incremental improvements.

Quality management is the methodology for achieving quality. Upper management gets involved in the quality council and in setting goals. Management needs to make the resources accessible that are necessary for the identified improvements. Management reviews the progress of the improvements and provides recognition and rewards where appropriate. Management needs to be the motivating voice behind the improvement process. This often includes stressing that employees will not lose their jobs as a result of any improvements that are implemented.

Quality planning focuses on customer needs and expectations, and then identifying the processes that need to be optimized in order to

Table 3.4 Timetable for Installing TQM

Activity	Months
Study the alternatives and formulate an annual quality improvement plan	6
Select a test site for conducting a pilot test (including training)	12
Evaluate the test site results; revise the approach as needed	18
Scale up across the firm and integrate TQM into the business plan	24

achieve those needs. In quality planning we identify what is important to the customer, which includes some level of market research, including customer interviews. Juran feels that statements like "quality first" are meaningless without a plan to move those actions forward.

Juran offers a quality planning timetable for installing his version of TQM. Table 3.4 shows his vision for this timetable.

For Juran, quality control is analyzing quality performance data and comparing this with the targets, identifying gaps, and taking corrective action. It uses a feedback loop by using sensors, like market surveys or SQC charts, and then reacting to the data collected by these sensors.

Self-control of quality is the motivation mechanism that gives workers ownership of the process and helps them respond to variations in the process. Self-control requires workers to

1. Have a knowledge of the goals and standards
2. Be capable of measuring and understanding performance
3. Be capable of changing the process to bring it into conformance

The key for self-control is "controllability." The worker must understand and be capable of working with the feedback loop and its sensors. Workers must have the capabilities in place to communicate their concerns to management. Other things that need to be in place include

▲ A communication mechanism between management and workers that works both ways
▲ Training options
▲ Management involvement in the quality process

The rules for measures that work well for effective quality include

▲ A specific unit of measure that permits evaluation
▲ A sensor that tracks and reports on the unit of measure on a timely basis that is responsive enough to take corrective action before it is too late
▲ Guidelines on how to respond to measure deviations

Juran stresses a series of errors that often occur in the perception of quality. These include the following:

▲ Management loses control of quality when the workers affect the quality process
▲ Quality "specialists" lose their role as experts
▲ Employees worry that improved quality may eliminate their job
▲ The workers are the problem when it comes to quality, not management
▲ Quality is all about attitude

Juran stresses that a clear understanding and use of self-control, the sensors, and the feedback loop are the key to TQM success.

1.2 TQC—Total Quality Control

Moving forward with our list of quality improvement techniques, and maintaining consistency with Table 3.1, we see in Table 3.5 that total quality control (TQC) is the next quality methodology listed.

There is a significant difference in focus between TQM and TQC, and the difference lies in the words "management" versus "control." TQM manages the quality process, whereas TQC controls the process. In this sense, TQC is a subset of TQM. However, since in the West TQC is utilized as an independent tool, separate from TQM, we need to list and define it separately.

TQC focuses on establishing a set of control points and control mechanisms where quality is checked. It uses mechanisms like SPC to monitor process performance and variability. Then, as the process departs from an acceptable range of performance, a feedback mechanism is initiated to inform an operator, or a manager, or both that corrective action is needed.

Table 3.5 Quality Methodology

1	TPS—Toyota Production Systems
1.2	TQC—Total Quality Control

TQC is a preliminary stage of what was later renamed Six Sigma. Six Sigma (6σ) incorporates the control pieces of TQC and builds a management structure around it, which we will see later in this chapter.

TQC focuses on being a quality watchdog, making sure that quality meets standards at specific points in the process. It does not focus on the management or motivation pieces of quality. It is, therefore, a system that is utilized by organizations that are in the early stages of incorporating quality into their corporate strategy.

1.3 Lean

Lean vs. Six Sigma vs. TOC

At this point we will explore a barrage of variations of TPS utilized as the tools of two major schools of thought: Lean and Six Sigma. Lean and Six Sigma each focus on two different aspects of the quality problem, much the same as the Deming, Juran, and Crosby variations of TQM. Lean focuses on a "systems" perspective and sees everything as a process flow. Six Sigma focuses on process variation and variability. Depending on the problem being addressed, each of these two methodologies has appropriate application. Unfortunately, there is a lot of contention between users of the two philosophies, much the same as there is between Deming, Juran, and Crosby adherents. Who is right depends on the problem being addressed. And sometimes none of them are right. Ironically, all three methods came from an attempt to copy the same master plan: the Toyota Production System (TPS). Incredibly, we see the same thing is true for Lean and Six Sigma. The harmony and cooperation that TPS promotes seem to have been lost when it comes to consulting fees.

The contention between Lean and Six Sigma also exists between the advocates of Lean, Six Sigma, and theory of constraints (TOC), which is not a TPS-based concept. Each adherent believes their tool is best and will give you all kinds of reasons why the other tool isn't as good. The reality is that all three of these tools have their appropriate need and fit. Each is better at solving certain types of problems better than the other. And they also cross-utilize each other's tools. For example, Lean uses Six Sigma as a statistical tool when quality process variation needs analysis. And Six Sigma uses value stream mapping (VSM) when process flow problems exist. Table 3.6 attempts to give an abbreviated comparison of the tools.

Table 3.6

	Lean	Six Sigma	TOC
Goal	Reduce waste, increase process speed, increase customer satisfaction	Improve performance on the customer's critical to quality (CTQ) items, reduce variation	Manage constraints
Tools	TPS tools	DMAIC and TQM tools	Bottleneck analysis
Focus	Process flow or system focused	Problem focused	System constraint focused
Method	Kaizen events, VSM	DMAIC and SPC	Constraint/ bottleneck analysis
Metric	Speed (cycle time), waste reduction, quality, customer satisfaction	Quality, cost	Increase throughput, reduce cost
Loop	Nine-step or PDCA	DMAIC	Constraint focused
Assumptions	Waste removal will improve business performance	A problem exists	Emphasis on speed and volume
Where Should It Be Used?	In process or flow environments, especially if we have a repetitive process	In data-intensive environments, for example, in quality defect analysis	In process flow environments that have a bottleneck that disrupts and limits throughput

The Lean Approach

A large number of tools are utilized as both Lean tools and Six Sigma tools. Rather than get involved in a debate about who really owns the tools, I have chosen to group all of the "commonly used" tools as Lean tools, realizing that Six Sigma may also have claim on them. Table 3.7 lists the Lean tools that we will be discussing in this section.

Table 3.7 Quality Methodology

1	TPS—Toyota Production Systems
1.3	Lean
1.3.1	Process, Project, or Event Charter
1.3.2	Scan
1.3.3	Kaizen Events, Rapid Improvement Events (RIE)
1.3.3.1	PDCA—Plan Do Check Act
1.3.3.2	A3 Reporting
1.3.3.2.01	Root Cause Analysis
1.3.3.3	Acceptance Change Management Tools
1.3.3.3.01	Change Acceleration Process (CAP) Model
1.3.3.3.02	Kotter Change Model
1.3.3.3.03	Myers Briggs
1.3.3.3.04	JoHari Window
1.3.3.3.05	Ladder of Inference
1.3.3.3.06	PAPT
1.3.3.3.07	Situational Leadership
1.3.3.4	Technical Tools
1.3.3.4.01	Value Stream Mapping
1.3.3.4.02	Seven Wastes
1.3.3.4.03	System Flow Chart
1.3.3.4.04	Swimlane Chart
1.3.3.4.05	Spaghetti Chart
1.3.3.4.06	5S
1.3.3.4.07	Cellular Work Design
1.3.3.4.08	JIT—Just-in-Time
1.3.3.4.09	Agile
1.3.3.4.10	Poka-Yoke
1.3.3.4.11	PQ Analysis
1.3.3.4.12	TPM—Total Product (Productive, Preventative or Production) Maintenance
1.3.3.4.13	Visual Workplace
1.3.3.4.14	DFM—Design for Manufacturability
1.3.3.4.15	SMED—Single Minute Exchange of Die, Quick Changeover
1.3.3.4.16	Kanban
1.3.3.4.17	Jidoka

1.3.3.4.18	Standard Work
1.3.3.4.19	Brainstorming
1.3.3.4.20	Fishbone Charting
1.3.3.4.21	Eight-Step Problem Solving
1.3.3.4.22	SWOT—Strengths, Weaknesses, Opportunities, Threats
1.3.3.4.23	VOC—Voice of the Customer
1.3.3.4.24	Gemba Walk—Go and See
1.3.3.4.25	Gap Analysis
1.3.3.4.26	Error Proofing/Mistake Proofing
1.3.3.4.27	SIPOC or COPIS
1.3.3.4.28	Pull Signaling
1.3.3.4.29	Affinity Diagrams

Lean has been an important tool toward facilitating the following improvements, among others:

▲ Eliminate waste
▲ Reduce cycle and flow time
▲ Increase capacity
▲ Reduce inventories
▲ Increase customer satisfaction
▲ Eliminate bottlenecks
▲ Improve communications

So what is "Lean"? As already mentioned, Lean is the Westernization of a Japanese concept that has carried several names. It's been known as the Toyota Production System, Just-in-Time (JIT), pull manufacturing, total quality management, and so on. Each of these names incorporates some aspect of Lean. What we know as "Lean" today isn't really any of these any more. One possible definition of Lean, taken from MainStream Management, a Lean consulting company, is

Lean is a systematic approach that focuses the entire enterprise on continuously improving quality, cost, delivery, and safety by seeking to eliminate waste, create flow, and increase the velocity of the system's ability to meet customer demand.

What we call "Lean" today is a collection of tools and methodologies, very few of which are actually "required" in any specific Lean process. When working on a specific Lean project, what is required is the design and assembly of the correct mix of tools to optimally facilitate the desired result. For example, if the goal is to improve the flow time for a process, we would assemble one set of tools, and an entirely different set of tools would be used if we were trying to improve quality.

Lean has developed into its own CPI environment, and along with that it has developed its own award process, the Shingo Prize for Excellence in Manufacturing (see Figure 3.2). In fact, the Shingo Award program has become the international standard for what Lean should look like. Therefore, as we look at defining Lean, it would be appropriate to start with the Shingo model.

The Shingo transformation model evaluates Lean performance in the following "dimensions":

1. Cultural enablers (respect for the individual/humility)
   a. Leadership and ethics: Implementing world class strategies and practices requires an enlightened leadership and organizational culture.
   b. People development: Respect for the individual and his or her development
      (1) Education, training, and coaching: People development and knowledge transfer
      (2) Empowerment and involvement: Is the organization fully capitalizing on the knowledge base?
      (3) Environmental and safety systems: Focus on the health and safety of the employees
2. CPI: Focused on the deployment of the appropriate tools and techniques
   a. Lean principles and ideas: Clarity and understanding of Lean principles. This section is quite extensive and goes through all the tools of Lean, including flow, pull, value, scientific thinking, and waste elimination.
   b. Value stream and support processes: This section reviews the areas where Lean can be applied. It checks the value streams in customer relations, product/service development, operations, supply, and management.

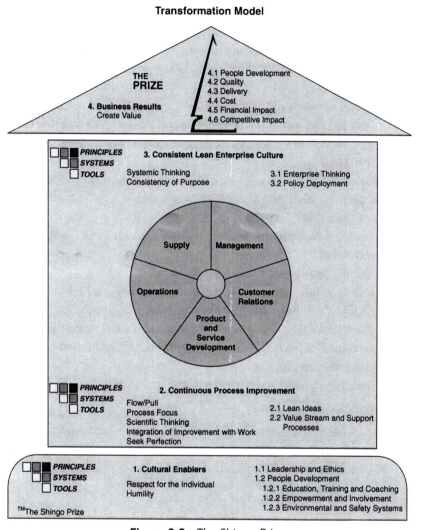

Figure 3.2   The Shingo Prize.

3. Consistent Lean enterprise culture: Here, the Lean practitioners are evaluating the degree of integration between manufacturing and non-manufacturing functions of the organization, and the extent to which Lean improvement tools have been applied in nonmanufacturing settings. They look for systemic thinking, holistic thinking, dynamic thinking, closed-loop thinking, and constancy of purpose.

a. Enterprise thinking: Here they make sure the full potential of Lean is realized throughout the enterprise, including areas like financial reporting, other reporting, business development, organizational design and development, information management, and leadership development.

b. Policy deployment: This section validates that the proper leadership is in place to instill the principles that will execute strategic plans. They look for scientific thinking, catch ball, and A3 thinking.

4. Business results: This is focused on creating and flowing value through the enterprise such that the customer is willing to pay for the product or service created.

a. People development: Protecting and fostering the growth of the individual.

b. Quality: Ensure that there are no errors.

c. Delivery: Ensure that the customer is getting what they want when they want it. Are customer expectations being met?

d. Cost: Continuous improvement focused on cost reduction.

e. Financial impact: Consistent and predictable growth in cash flow compared to risk.

f. Competitive impact: Long-term growth, market share growth.[1]

The Shingo Prize organization uses these criteria to evaluate prize applicants. These criteria are extremely valuable in the evaluation of any organization's lean and continuous improvement performance.

Some of the key principles behind Lean include the following:

1. Define "value" from the customer's perspective.
2. Define the process by looking at and analyzing all the pieces of the Supplier—Input—Process—Output—Customer (SIPOC). This should define the process value stream and identify opportunities for waste elimination.
3. Remove obstacles (bottlenecks) that disrupt the value flow.
4. Drive product and service flow at the "pull" of the customer.

---

[1] Utah State University, College of Business, *The Shingo Prize for Operational Excellence—Model—Application Guidelines*, 3rd Edition 2008, www.shingoprize.org.

5. Empower employees in the change process through teaming.
6. Build a strategic plan that focuses on the goals of the enterprise and that focuses on continuous improvement. Update the plan annually. Use this plan to identify strategic targeted areas of improvement.

Lean is primarily about team building, integration, and ownership. A Lean facilitator is tasked with organizing the appropriate teams and then giving them the appropriate guidance and training in the selected tools so that the Lean effort can progress with the greatest efficiency. The Lean team is the one that makes the decisions about any changes in process, and they have ownership of these changes. It is the role of the facilitator to help the team stay on task as they develop and implement these changes in the minimal amount of time.

There are three fundamental stages to any successful Lean process: the acceptance stage, the technical stage, and the sustainment stage. We start with the acceptance stage, where we analyze the organization's recognition of the need for change.

ACCEPTANCE STAGE   The first thing we need in any Lean process is a trained Lean facilitator who is highly experienced in the process and its available tools. This facilitator will initially talk to the Champion of the organization and get guidance on where leadership wants him or her to focus their efforts. Since Lean is disruptive in that it draws away a large number of resources, there has to be a compelling reason to do Lean. This compelling need must come from the organization's strategic priorities. For example, the need may come from the company's position in their respective industry and their need to be able to adapt to changes that might affect their growth and survival. The facilitator needs a clear, measurable objective for his Lean activity.

The facilitator may next choose to perform a scan of the team members. This involves holding short 15- to 30-minute face-to-face meetings with each of the management team members to get their perspective on the upcoming event. The information from this scan is only for the eyes of the facilitator, and the purpose of the scan is to identify the key areas of concern. The facilitator now knows all the key areas of the organization, all customers, all suppliers, and any other stakeholders that need to participate in the rapid improvement event (RIE).

Once the team has been identified and invited, some of the acceptance tools are used to evaluate the change readiness and the dynamics of the team. All this is done before the team meets so that the facilitator can be ready for the team that he or she will be encountering.

Now we are ready to get the team together in the same room. We start with a training session, teaching them the whys and hows of Lean. Then we show them what their role will be in the Lean process.

The facilitator is not the leader of the Lean effort. The facilitator should subordinate himself or herself to the Champion and team leader, and will take direction from them on what needs to be done, when it needs to be completed, and how it should be done. The process owners are the team, and they need to own the changes to the process. The facilitator needs to give the team direction in the form of training about the Lean tools. And the facilitator needs to share findings, problems, and successes with the team.

TECHNICAL STAGE   The technical stage begins with the event, which, almost always, is some type of process mapping activity where the team tries to thoroughly understand the current state process or system under study. The primary tool for this is almost always a value stream mapping (VSM) exercise, where the current "value stream" is mapped out in detail (CS-VSM). There are also other tools, like spaghetti charting, which focuses on the people movement, or systems flow charting, which focuses more specifically on the information flow of the organization. The objective of all these maps and charts is to study the process in as much detail as possible so that waste, bottlenecks, systems holes, and other opportunities for improvement can be identified.

After the technical assessment, the findings are used to develop improvement task lists. Each event creates a task list, also referred to as an action item list or a Lean newspaper. This action item list is then moved forward until each action item is resolved.

SUSTAINMENT STAGE   The sustainment stage is where the Lean effort takes on a life of its own. At this point everyone has been trained at all levels of the organization and they now have taken ownership of their piece of the process. The Lean process no longer needs the facilitator. The Champions meet regularly and give direction to each of the event RIE team leads. RIE teams are organized as new areas for improvement are identified, and obsolete RIE teams are disbanded as they achieve their desired goals. Lean becomes a way of life for the organization as a whole.

1.3.1   Process, Project, or Event Charter

The charter is a contract between a project Champion(s), a team leader, the project team, and the dollar resources that will be needed. A charter should be generated to make sure you have a team and the appropriate leadership support to implement the task list that will be generated. Each of these tasks may also require a charter to make sure appropriate meaningful support exists so that you can complete this activity.

The project charter defines the project, identifies who is participating in the project, and ties down the timing of the project. Information that is generally included in the charter is

- Project/process name
- Champion—The person with the money to see this through/someone who believes in the potential benefits of this improvement process
- Process owner whose workplace will be directly affected by this change
    - What are all the areas that will be impacted by this process change?
- Product/service being impacted
- Team leader—Assigned by the Champion
- Facilitators—Not necessarily experienced with the project or process, but highly experienced in the Lean process
- Dates (actual and estimated)
    - Project start date
    - Measurement completed
    - Preanalysis completed
    - Improvements (tasks) completed
    - Project completion date
- Expected savings/benefits
    - What are the key performance indicators (KPI) and what is the anticipated improvement goal for each
    - Who owns the metric
    - Benefits to the external customer
- Rapid improvement event (RIE) or project dates and time
- Linkage to the strategic roadmap—The project needs to tie directly to the organization's strategy
- One-paragraph explanation of the improvement opportunity
- One-paragraph description of the desired outcomes/results
- Project deliverables

▲ Project scope—Where does the process start and end (as far as this improvement process is concerned)?

▲ Team members/where assigned (what office do they work in?)/contact information (phone and e-mail)

▲ Follow-up plan

    ▼ Executive out-brief of the results of the exercise to the Champion and the process owners

    ▼ Follow up meetings—How often will we have follow-up status meetings? Who should attend?

▲ Approvals

    ▼ Champion approval signature and date

    ▼ Process owner approval signature and date

    ▼ Executive out-brief date

Filling out the charter and getting the appropriate signatures is critical in making sure that everyone is on board with the planned activities. Without a signed-off charter you may find yourself spinning your wheels when it comes time to allocate funds on the suggested improvements.

1.3.2 Scan

The objective of the scan is to get a quick, high-level read on current operating conditions. The scan focuses on interviewing top management in order to get a read on their goals and objectives. In the scan we want to clearly define the target area that is under consideration for quality improvement. Then we try to define the purpose and function of the targeted areas. The project facilitator will interview management, looking for management's perception of problem areas, like inconsistent or intermittent flow, quality failures, or bottlenecks. The facilitator records current problems, concerns, and opportunities for improvement on a checklist. The focus of the scan is to give direction to the Lean event team. The scan makes sure we are on track with management expectations when we go into an event, and then the Lean facilitator can monitor the progress of the team to make sure that the team stays on track during the actual event.

During the scan we photograph, collect data about, and document problem areas and put them on a display board so the entire team can review and evaluate these areas when it comes time to execute the event. Later, during the event, the team searches for alternative solutions to these problems. For example, is the copy machine or the printer too far away from the workplace, resulting in excessive travel time "waste"? Sometimes we "centralize" equipment of this type so that it is easier to control the equipment, but the result is that we cause our employees to "waste" hundreds of hours in travel time going back and forth to these central locations.

1.3.3   Kaizen Events and Rapid Improvement Events

The Lean event, also referred to as an RIE or kaizen event, is a structured process for implementing change. The typical Lean event follows a structured sequence of steps and is managed by the Lean facilitator; however, there is no "typical Lean event." Every event is customized by the Lean facilitator to maximize a particular improvement process. Typical event steps could include the following:

- Scan top management
- Develop a nine-step A3, which outlines and gives structure to the quality improvement process
- Define the team charter—From this we get top management/Champion support and we identify the team that will be used for the event
- Perform team readiness and change readiness assessment to make sure you have a balanced team
- Schedule the event (usually about one concentrated week when everyone focuses on the improvement process)
- During the actual one-week event, we typically do the following:
  - Lean training, which at a minimum should include:
    - Explanation of what Lean is (definition)
    - Discussion of the Seven Wastes
    - Explanation of the SIPOC and SWAT tools
    - Explanation of the VSM process
    - Review of the nine-step A3 document and how it is the guideline for the Lean event
  - Review and perform a Strengths—Weaknesses—Opportunities—Threats (SWOT)
  - Review and perform a Supplier—Input—Process—Output—Customer (SIPOC)
  - Current-state VSM
  - Ideal-state VSM
  - Future-state VSM
  - Improvement task list

▼ Execution plan for the task list
▼ Plan a "governance process," which is where the team meets regularly to do a status check on the improvement process
▼ Report to the Champion
▲ Hold regular follow-up status (governance) meetings

The next subsections will focus on the quality improvement elements and tools that may be used during a Lean RIE.

1.3.3.1 PDCA—Plan Do Check Act

The Loops

CPI problem solving is a series of never-ending loops. We are never done. There is always something that can be improved upon. In addition, the environment is constantly changing. Circumstances change. Customers change. Technologies change. And as long as something is changing there will be opportunities for improving what was changed.

If you search the change management literature, you'll find never-ending loops. We have the Lean PDCA (Plan-Do-Check-Act) loop, also referred to as Deming's Quality Wheel (see Figure 3.3); we have the Six Sigma DMAIC (Define, Measure, Analyze, Improve, Control) loop; and we have the TOC (theory of constraints) loop. With a little analysis we quickly learn that all

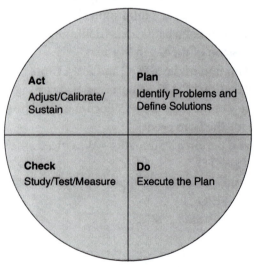

Figure 3.3 Deming's Quality Wheel—PDCA.

Table 3.8   "Loop" Comparison

Basic Steps of Any "Loop"	PDCA	DMAIC	TOC
Step 1—Clarify and Validate the Problem	Plan	Define	Identify the constraint
Step 2—Perform a Purpose Expansion on the Problem (Breakthrough Thinking)		Measure	
Step 3—Break Down the Problem/ Identify Performance Gaps			
Step 4—Set Improvement Targets			
Step 5—Determine Root Cause		Analyze	
Step 6—Develop Improvement Task List	Do		Exploit the constraint
Step 7—Execute Improvement Tasks		Improve	Subordinate the processes
Step 8—Confirm Results	Check	Control	Repeat the process—find another constraint
Step 9—Standardize Successful Processes	Act		

these loops are basically the same. Table 3.8 shows how each of these tools parallels the other in the change management cycle. All of these loops are variations of the TPS, which is the foundation of Lean and Six Sigma.

1.3.3.2   A3 Reporting

One of the foundational tools for TPS (or Lean) is the A3 report. This tool asks the question, "Why are we even doing this project at all?" Using the A3 in the TPS environment we see multimillion dollar projects approved or rejected based on the information on this one simple tool. It is referred to as the A3 report because of the size of the paper that the report is created on. The entire report is on one sheet of paper. It's not presented in a 200-page document, or a 100-slide "death by PowerPoint" deck. The entire analysis is performed and reported on one sheet of A3-sized paper. But it's not the size of the paper that's important; it's how the paper is organized and what information it contains. The A3 tool becomes a critical tool to make sure you're not wasting time; that you're doing the right things.

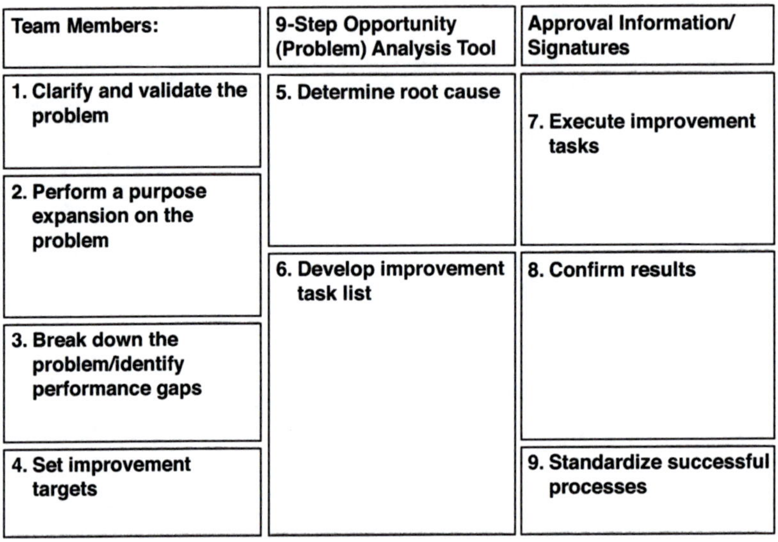

Team Members:	9-Step Opportunity (Problem) Analysis Tool	Approval Information/ Signatures
1. Clarify and validate the problem	5. Determine root cause	7. Execute improvement tasks
2. Perform a purpose expansion on the problem		
	6. Develop improvement task list	8. Confirm results
3. Break down the problem/identify performance gaps		
4. Set improvement targets		9. Standardize successful processes

Figure 3.4 9-Step A3.

The A3 example outlined here is not the only A3 format available. There are many others, and you can find them in a variety of Lean books. However, the author has experienced a great deal of success using the tool presented here for both opportunity analysis (the first seven steps) and later for reporting progress (steps 8 and 9).

In the A3 that this book is recommending there are nine steps. The report itself looks much like the example in Figure 3.4. This same format is used for opportunity analysis, project presentation, project justification, and project performance review. When you use the same format consistently, the entire team (management and users) become familiar with it. Everyone knows what they are looking for and they go right to the box that interests them.

1.3.3.2.01 Root Cause Analysis

Root cause analysis is step 5 in Figure 3.4. In this we are defining the root causes of the current problem and the reasons for the current performance gaps. We try to answer the question, "What caused the need for this change?" We look to find out what is happening or not happening that is creating the problem. Here are some questions that should be resolved:

1. What root cause analysis tools are necessary?
   a. Why are these tools necessary?

    b. What benefit will be gained by using them?

    c. Who needs to be involved in the root cause analysis?

        (1) Ten heads are better than one

        (2) Remember "cultural" issues related to the problem

2. What are the root causes according to the tools we used?

3. How will the root causes be addressed?

4. Will addressing these root causes address the performance gap?

5. Can the problem we are trying to solve be turned on or off by addressing the root cause?

6. Does the root cause make sense if the five Whys are worked in reverse?

    a. Working in reverse, say "therefore" between each of the "whys."

7. Root cause analysis results should describe a clear and coherent cause-effect chain that demonstrates an in-depth understanding of the problem in context, showing how the root cause is linked to the observed problem. Root causes may result from a poorly specified activity, an unclear connection, or a complicated or undefined pathway. Tools that are available to help in this analysis include:

    a. Five Whys—Reiterative process of asking "why" to go beyond superficial analysis. Start with "Why is this problem occurring?" Then, using that answer, ask "Why?"And iteratively repeat this process five times to see what this teaches us about the cause of the problem.

    b. Brainstorming—This is a simple tool where we get several people into a room and start writing on a chalk board or a white board and start asking "Why?" to the group to see what they come up with.

    c. Fishbone charts—A fishbone chart is a tool that is used to organize brainstorming ideas into categories.

    d. Pareto charts—A Pareto chart tries to identify those factors with the greatest impact on the problem using bar charts.

    e. Affinity diagrams—The affinity diagram is an alternative way to group, prioritize, and organize the results of the brainstorming exercise.

    f. Control charts—Control charts are a statistical tool that analyzes performance data in an attempt to identify out-of-control processes. Control charts are used over a period of time to monitor performance and to track data movement.

1.3.3.3 Acceptance Change Management Tools

The acceptance stage of a Lean event is the first of three fundamental stages to any successful Lean process (the others are the technical stage and the sustainment stage). The acceptance stage is where we analyze the organization's recognition of the need for change. Several acceptance tools are used to evaluate the change readiness and the dynamics of the Lean team. All this is done before the team meets so that the facilitator can be ready for the team that he or she will be encountering.

1.3.3.3.01 Change Acceleration Process (CAP) Model

MainStream Management, a leading Lean consulting company, uses an adaptation of the change acceleration process (CAP) model (see Figure 3.5) for their change model. This was adapted from the model used by GE, which was originally developed by Noel Tichy and which is similar to Kotters change model. The model indicates several stages of change in the change management process. The first section is about selling the Lean

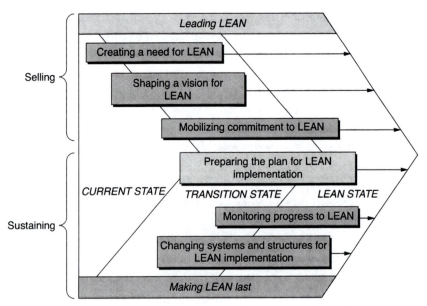

Source: Adapted from the model developed for GE by Noel Tichy, similar to Kotters Change model.

Figure 3.5 Change Acceleration Process (CAP Model).

change process to the organization. The second section focuses on sustaining the Lean change process so that the process continues even after the facilitator discontinues his oversight activities.

In the selling phase of the CAP change model we see the creation of a need for Lean within the organization. This is accomplished by identifying a need for change. We ask questions like: What are some large problem areas, or strong growth needs, of the organization? Without the proper change management driver, the problem may never get resolved or, worse yet, it may not get resolved in the best possible way. We need to identify opportunities where the Lean change process can be effective.

Shaping the vision for Lean requires us to look first at the goals of the organization and then to look for ways to help the organization optimize those goals. Mobilizing the commitment to Lean requires team identification and team building. Preparing a plan requires timelines and schedules. It requires a plan of attack with specific steps. Monitoring the Lean process requires checkpoints and measures.

Change systems and structures for Lean implementation require ownership and commitment from top management down to the workers on the floor. The teams need to own the change process and need to be excited about it so that they will maintain it even after the facilitator leaves. This is about putting structures and systems in place that continue to facilitate the Lean change management process.

1.3.3.3.02   Kotter Change Model

Another change model that can be used is the one designed by Kotter (see Table 3.9). This model can be used in place of the CAP model. It contains an eight-stage process for implementing change.

1.3.3.3.03   Myers Briggs

If the Lean facilitator wants to look deeper into the personalities of the team members, a good tool for this evaluation is the Myers Briggs assessment. This tool was developed in the 1950s and has proven itself an excellent tool for helping people understand themselves. It ranks everyone as follows:

- How we energize ourselves (in the range from E to I)
  - Extravert (E)
    - Attention seems to flow out to objects and people in the environment

Table 3.9   The Eight-Stage Process of Creating Major Change

---

1 Establishing a sense of urgency
- Examining the market and competitive realities
- Identifying and discussing crises, potential crises, or major opportunities

2 Creating the guiding coalition
- Putting together a group with enough power to lead the change
- Getting the group to work together like a team

3 Developing a vision and a strategy
- Creating a vision to help direct the change effort
- Developing strategies for achieving that vision

4 Communicating the change vision
- Using every vehicle possible to constantly communicate the new vision and strategies
- Having the guiding coalition model the behavior expected of employees

5 Empowering broad-based action
- Getting rid of obstacles
- Changing systems or structures that undermine the change vision
- Encouraging risk taking and nontraditional ideas, activities, and actions

6 Generating short-term wins
- Planning for visible improvements in performance, or "wins"
- Creating those wins
- Visibly recognizing and rewarding people who made the wins possible

7 Consolidating gains and producing more change
- Using increased credibility to change all systems, structures, and policies that don't fit together and don't fit the transformation vision
- Hiring, promoting, and developing people who can implement the change vision
- Reinvigorating the process with new projects, themes, and change agents

8 Anchoring new approaches in the culture
- Creating better performance through customer- and productivity-oriented behavior, more and better leadership, and more effective management
- Articulating the connections between new behaviors and organizational success
- Developing means to ensure leadership development and succession

- ▼ Desire to act on the environment
- ▼ Takes action
- ▼ Impulsive, frank
- ▼ Communicates easily
- ▼ Sociable
- ▼ Introvert (I)
  - ▼ Attention to the inner world of concepts and ideas
  - ▼ Reliance on enduring concepts versus transitory external events
  - ▼ Thoughtful, contemplative detachment
  - ▼ Enjoyment of solitude and privacy
- ▲ What we pay attention to (in the range from S to N)
  - ▼ Sensing (S)
    - ▼ Focus on here and now, immediate situation
    - ▼ Enjoys the present moment
    - ▼ Realistic
    - ▼ Acute powers of observation
    - ▼ Memory for details
    - ▼ Practical
  - ▼ Intuition (N)
    - ▼ Tie seemingly unrelated events together
    - ▼ Creative discovery
    - ▼ Perceive beyond the here and now
    - ▼ May overlook current facts
    - ▼ Theoretical, abstract
    - ▼ Future oriented
- ▲ How we make decisions (in the range from T to F)
  - ▼ Thinking (T)
    - ▼ Tough minded
    - ▼ Makes logical connections
    - ▼ Principles of cause and effect
    - ▼ Tends to be impersonal
    - ▼ Analytical, objective
    - ▼ Concern for justice
    - ▼ Desires a connection from the past to the present to the future
  - ▼ Feeling (F)
    - ▼ Tender minded

▼ Weighs the relative merits of the issues
▼ Applies personal and group values subjectively
▼ Attends to what matters to others
▼ Concern for the human as opposed to the technical aspects of the problem
▲ How we are oriented to the world (in the range from P to J)
  ▼ Perceptive (P)
    ▼ Open to incoming information
    ▼ Curious and interested
    ▼ Spontaneous and curious
    ▼ Adaptable
    ▼ Open to new events and changes
    ▼ Willing to take in more information before making a decision
  ▼ Judging (J)
    ▼ Seeks closure
    ▼ Organizes events
    ▼ Plans operations
    ▼ Shuts off perceptions as soon as they have observed enough to make a decision
    ▼ Organized
    ▼ Purposeful
    ▼ Decisive

This analysis can be invaluable in team creation because what you need on your team is a balance. Too many of any one personality type can disrupt the decision-making process. In Figure 3.6 we can see all the different categories. If, for example, we had a lot if I's and few or no E's on the team, we would have limited and poor group discussions.

In the end, Myers Briggs assessments help both the facilitator and the team members. They

▲ Help individuals to know themselves
▲ Give a tool to "be with" others
▲ Give a baseline from which to develop adaptive behaviors
▲ Help people learn how to better communicate by understanding and by talking with others in their "language"
▲ Develop better facilitation methods
▲ Develop more effective exercises

ISTJ Analytical MANAGERS of FACTS/DETAIL	**ISFJ** Sympathetic MANAGERS of FACTS/DETAIL	**INFJ** People Oriented INNOVATORS	**INTJ** Logical, critical, decisive INNOVATORS
ISTP Practical ANALYZER	**ISFP** Observant, loyal HELPER	**INFP** Imaginative, independent HELPER	**INTP** Inquisitive ANALYZER
ESTP REALISTIC ADAPTERS material world	**ESFP** REALISTIC ADAPTERS human relations	**ENFP** Warm, enthusiastic PLANNERS of CHANGE	**ENTP** Analytical PLANNERS of CHANGE
ESTJ Fact-minded practical ORGANIZER	**ESFJ** Practical HARMONIZER	**ENFJ** Imaginative HARMONIZER	**ENTJ** Intuitive, innovative ORGANIZER

Figure 3.6   Myers Briggs type indicator.

⏶ Appreciate the differences in others
⏶ Teach us the value of type diversity
⏶ Learn to accept others for who they are
⏶ Leverage each person's type by identifying their role in a team environment
⏶ Conduct more effective meetings
⏶ Help understand how previously annoying behavior can be seen as amusing, interesting, and as a strength

The Web sites for the Myers Briggs assessments are http://www
.humanmetrics.com/cgi-win/jungtype.htm or http://www.myersbriggs
.org/my_mbti_personality_type, which introduces you to the Myers
Briggs test. These lead you to the website http://www.humanmetrics.com/
cgi-win/JTypes1.htm where you can actually take the test and get your
score.

1.3.3.3.04 JoHari Window

The JoHari window is a tool that offers insight into how we see ourselves and how others see us. It helps investigate the openness of team members and helps us understand team members' willingness to communicate. To accomplish this, the JoHari window uses "disclosure" (or telling) and "feedback" (or asking) to determine our individual openness. In the JoHari window we take a survey and then plot the results of this survey on a graph. The graph divides personalities into four segments, as seen in Figure 3.7.

The JoHari window describes, evaluates, and predicts interpersonal communication. The window panes show us how we present and receive information about others and ourselves. The size and shape of the panes will change over time. Using the JoHari segments we identify each of these segments as follows:

A (arena)   This represents the part of yourself that is known by you and also known by others. This window pane represents free and open exchange of information between others and yourself. This is public behavior information, which is available to everyone. The pane increases in size as the level of trust increases between others and yourself. As more information, particularly personally relevant information, is shared, this trust increases. This represents a manager with a capacity for open relationships. In this window the

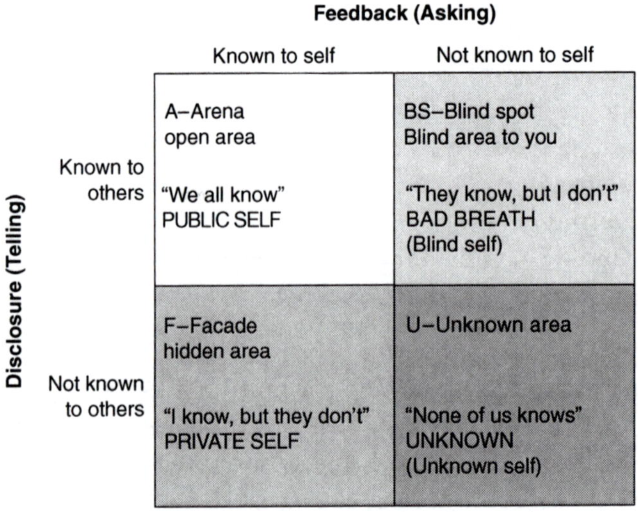

Figure 3.7 Jo-Hari window.

glass is two-way. There is an open exchange of facts, feelings, and opinions between the people communicating through this pane.

BS (blindspot)   This represents what others know about you, but that you do not realize about yourself.

F (façade)   This represents what is known by you about yourself, but is hidden from others.

U (unknown)   This is that part of you that is unknown by others as well as by yourself.

The goal of a good communicator is to enlarge the arena window as much as possible in a balanced fashion. To accomplish this, we attempt to reduce the "unknown" regions. We attempt to reduce what we don't know about ourselves and what others don't know about us. The result should be something like Figure 3.8.

We increase disclosure because we increase trust. We are more willing to give out information about ourselves. We share more of what we think and how we feel.

1.3.3.3.05   Ladder of Inference

The ladder of inference is a tool designed to help the user get a better grasp on the information that is being exchanged. It helps process information,

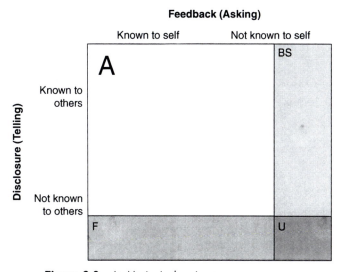

Figure 3.8   Jo-Hari window increases openness.

including thoughts and ideas, and assists us in arriving at meaningful con-clusions so that we take the most appropriate actions. The tool is used to help us avoid jumping to early conclusions that are often inaccurate. We strive to gather as much information as possible before formulating our conclusions. The ladder of inference promotes asking questions and shar-ing information in open conversations in order to make sure that we are on the right track toward an appropriate conclusion (see Figure 3.9).

The ladder of inference is a reasoning process that allows individuals to perceive events that surround them throughout their entire life. The inter-actions that occur during our lifetime affect the conclusions that we draw in our current and future encounters. The encounters are filters through which we see the world and they result in the opinions or beliefs that

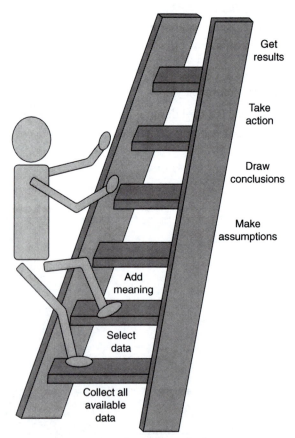

Figure 3.9 The ladder of inference.

we form about what we see and hear, and then they affect our reactions. In the ladder of inference we are cycling new data through our filters, thereby allowing us to formulate the most objective conclusions possible. Therefore, the objective is to gather as much data as possible and to defer judgment for as long as possible.

For the ladder of inference to be the most effective, we must become the masters of two activities: asking and telling. Asking is where we try to create an openness that allows others to express their opinions without fear of retaliation, thereby bringing as much information as possible to the surface. Telling is where we state our viewpoint in a tactful manner, which avoids conflict but still allows the listener to fully understand what we are saying. We climb the ladder using "telling" and we come back down the ladder by "asking" in an attempt to maximize the strength of our conclusions.

The first step of the ladder is to collect all the available data. The second step is to select the data. Not all data is relevant to the topic under consideration. The third step is to add meaning. Each person who views the data will add their own perspective and meaning to the information based on their personal feelings, passions, biases, and experiences. It may become necessary to validate the information so that uniformity of interpretation can be achieved. We need to carefully listen to viewpoints that are different from our own.

The fourth step is to make assumptions. The fifth step is to draw conclusions. The sixth step is to take action. The seventh step is to get results. Here we evaluate the results against the conclusions and assumptions we made as we climbed the ladder.

With an understanding of each of the steps in the ladder of inference, we can now see how the asking and telling process works. The process of asking and telling involves the use of correct body language and tone of voice. If done incorrectly, this may cause other stakeholders or team members to misinterpret your intentions and may cause the flow of information to be biased.

The flow up and down the ladder of inference using asking and telling techniques should become spontaneous and should be a natural part of the communication process. The ladder of inference takes practice and requires a good facilitator so that the users of the ladder learn how to move up and down this ladder naturally. Rushing up the ladder too quickly will result in taking inappropriate actions and thereby achieving undesired results.

1.3.3.3.06 PAPT

Another invaluable self-awareness and team-awareness mechanism is the PAPT test. This test focuses on the Passive-Aggressive (PA) dimension and compares it to the People or Task (PT) Oriented dimension. Each individual is ranked on these two dimensions, and from this ranking we can develop some conclusions about the dynamics of our team.

The PAPT process begins with a survey, and from the results of the survey we are able to chart the personality dimensions of each member of the team. In Figure 3.10 we see how the rankings occur graphically. In Figure 3.11 we see the significance and meaning of each of the quadrants.

The first quadrant is the Analyzer. This is an individual who is obsessed with getting the task correct. This is someone who wants to know as much as possible before making a decision.

The second quadrant is the Ruler. The third quadrant is the Relater. The focus this time is on helping everyone get along. The fourth quadrant is the Entertainer. This individual wants recognition and wants as big an audience as possible.

The test can be found in the book *Reinventing Lean: Introducing Lean Management into the Supply Chain* by Gerhard Plenert (Butterworth-Heinemann, 2006).

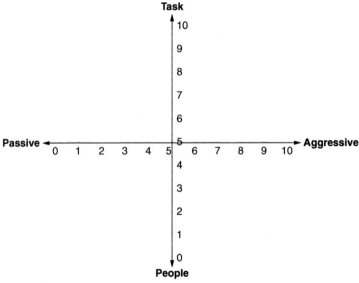

Figure 3.10 PA-PT model.

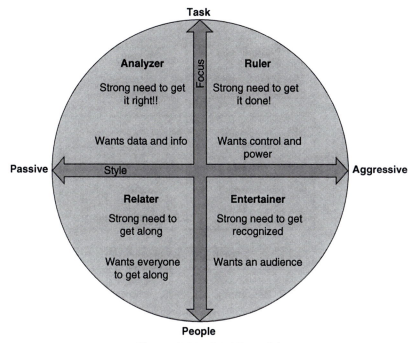

Figure 3.11   PA-PT model.

1.3.3.3.07   Situational Leadership

Situational leadership is a tool that attempts to help influence someone to complete a task successfully. We start with an understanding of the follower's needs. Does the follower want to complete the task? Does the follower know how to complete the task?

The leader's responsibility is to offer moral support in the form of reassurance. In addition, the leader offers direction by facilitating skills training and making the necessary resources available to complete the task.

The situational leadership follower grid is shown in Figure 3.12. Here we see four quadrants, each representing a different role. For D1 we have the enthusiastic beginner where we have a high level of commitment but a low level of competency. For the D2 individual we have the disillusioned learner. This is someone who is moderately eager and has an average desire to achieve. For D3 we have the reluctant contributor who is someone who is not eager at all. This individual has a low desire to achieve and is not

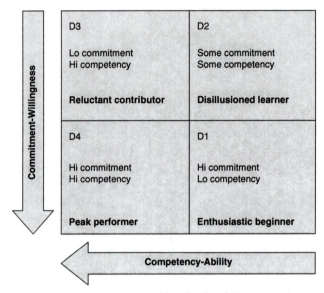

Figure 3.12 Situational leadership follower grid.

dedicated to the job or organization. In D4 we find the peak performer who is someone who is very eager and has a high desire to achieve.

The situational leadership leader grid is shown in Figure 3.13. Here we also see four quadrants, each with their own classifications. For S1 we have the directing/telling type of personality where the leader sets all the goals and solves all the problems. The leader controls the "who, what, when, where, how, why, and how much" of all decisions. This is one-directional, top-down communication where the leader supervises all and evaluates everyone and everything.

The S2 type of leader is the coaching/selling type of leader where again the leader sets all the goals, but in this case the leader consults with the follower. In the S3 leader we find a someone who is supporting and participating. The S4 type of leader is the delegator who follows defined problems and sets the way for participative problem solving.

The purpose of the situational leadership questionnaire is to discover your leadership style in a variety of situations. It tests your ability to recognize situations, and it evaluates your leadership effectiveness. It also helps facilitators determine what types of management they are dealing with. It helps them to determine a strategy for moving the project forward.

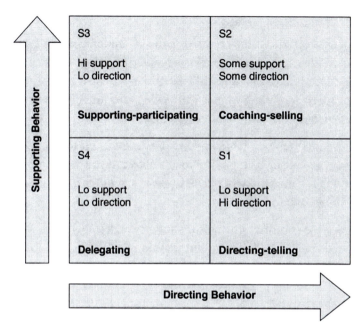

Figure 3.13   Situational leadership grid.

The details behind this process can be found in the book *Reinventing Lean; Introducing Lean Management into the Supply Chain* by Gerhard Plenert (Butterworth-Heinemann, 2006).

1.3.3.4   Technical Tools

The technical stage of a Lean process utilizes a large variety of tools. Which tool is used depends on the needs of the improvement process that is being executed. The primary tool for this is almost always a VSM exercise, where the current "value stream" is mapped out in detail (CS-VSM). This is usually followed by a future state VSM (FS-VSM) and a gap analysis. The objective of all the maps is to study the process in as much detail as possible so that waste, bottlenecks, systems holes, and other opportunities for improvement can be identified.

After the technical assessment, the findings are used to develop improvement task lists. Each event creates a task list, also referred to as an action item list or a Lean newspaper. This action item list is then moved forward until each action item is resolved.

1.3.3.4.01 Value Stream Mapping

The primary Lean tool used to assist teams in understanding a process is VSM. The value stream includes all the events and activities in a product or process's supply chain that affect the customer's perceived "value."

To properly use VSM we need to understand the difference between value-added and non-value-added activities.

▲ Value-added activity Anything that directly increases the value of the product or service being performed (from a customer's perspective).

▲ Non-value-added activity Any support activity that does not directly add value to the product or service.

In a service environment, "value-added" refers to the service. "Non-value-added" includes all the support functions that occur and that prepare the environment to perform the service.

Mapping the Process VSM is a waste identification tool that is used to identify Lean improvement opportunities by identifying the non-value-added processes. It does not focus on working harder; it focuses on working smarter.

VSM creates a visual display of the value stream by looking at the entire system, including all inputs, the process, and all outputs. The VSM shows the linkages throughout the system and challenges the current state of all activities. It identifies the sources of non-value-added waste.

In the end, VSM becomes the foundation for the development of an improvement plan. It presents a complete picture of the current state of the processes involved. Using this information, VSM can then establish a vision of what the desired future state should look like. This future state is used to develop the improvement plan, and from this we develop a collection of improvement actions.

The VSM process has four phases:

▲ Preparation
▲ Current state map
▲ Future state map
▲ Improvement plan

PREPARATION In the preparation phase we focus on identifying the scope of the system under consideration. We look for those processes that have the largest impact on the business. The product or service selected should be one that has an established a history of failures. The VSM should focus on a product that has a need for improvement.

Once the product or process has been selected, a team is organized, with a focus on the resident experts in that area. The team also needs to include a customer and supplier to the process. The team needs a facilitator, who is often referred to as the value stream focal. This individual makes sure that the VSM analysis process stays on track. This individual becomes the scheduler, negotiator, and arranger, making sure the VSM team has the resources they need.

MAPPING PROCESS   Before the actual mapping process can begin, the facilitator needs to make sure that the team has been properly formed to include all relevant stakeholders and product authorities. The team should include at least one process expert for every step in the product's value stream.

With the team in place, the facilitator organizes training on the VSM tools that will be used for the mapping exercise. Then the team goes on to define "value" from the perspective of the customer. This needs to be documented and displayed for the team and kept at the forefront of their VSM effort.

With the customer information documented, the team needs to review the key performance indicators (metrics) that are in place and then needs to challenge whether they are the correct performance indicators. These measures need to focus on the customer while at the same time balancing the needs of the other stakeholders with those of the members of the organization.

At this point, the team should be in place and the goals established. Now the team is ready to observe the process and to gather data on it. This often requires a detailed walk through the value stream. After the walk the team is now ready to create the VSM. An example of a very short process map is shown in Figure 3.14. Most maps will be much larger because the number of steps in a process can be quite extensive.

The VSM is reviewed to find the sources of waste (the high non-value-added times). With Figure 3.14 we see an enormous delay between data input and specs. This delay of 17.1 days also results in a large interim inventory of work, assuming that people are not just sitting around for the 17 days.

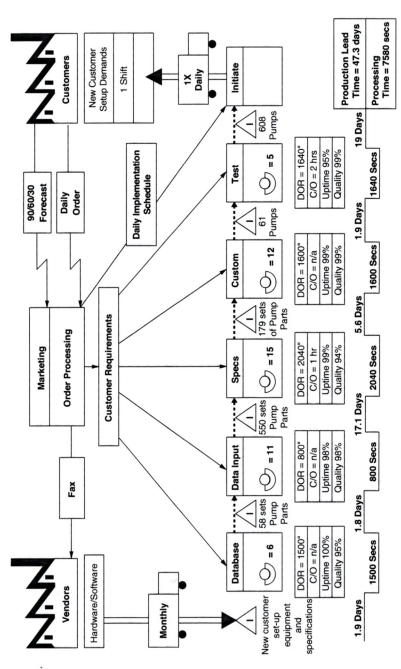

Figure 3.14   Sample value stream map (VSM)

Next in the mapping exercise comes the process of identifying improvements. This has the following steps:

- Create an ideal state VSM
- Use the ideal state VSM to create a realistic future state VSM
- Develop an action item list of improvement opportunities
- Classify/prioritize the action item list
- Select improvement events based on the highest priority areas of improvement

IDEAL STATE VALUE STREAM MAP/FUTURE STATE VALUE STREAM MAP   The objective behind developing a future state value stream map (FS-VSM) is to identify a target goal for our improvement effort. We start by ignoring the current state VSM and we redraw an ideal VSM (IS-VSM). How would the perfect system operate? Next, using the ideal VSM, we come down to reality, acknowledging that we have limited resources. This is an important thought process. We use our IS-VSM as the long-term goal when we create our FS-VSM. From this we can take our CS-VSM and identify changes that will need to be made in order to bring our current operating state to the desired future state.

Using the CS-VSM in Figure 3.14, we develop an IS-VSM and from that an FS-VSM, which can be seen in Figure 3.15. The differences between Figure 3.14 and Figure 3.15 define the improvement gaps. These are the opportunities for improvement, and the process of identifying these gaps is often referred to as gap analysis.

DEVELOP AN ACTION ITEM LIST OF IMPROVEMENT OPPORTUNITIES   Using Figure 3.14 as the current state and Figure 3.15 as the future state, we can now generate an action item list of improvements. In some Lean activities, this list is referred to as the Lean newspaper.

1.3.3.4.02   Seven Wastes

Taiichi Ohno developed the original catalog of the wastes and made them part of the Toyota Production System (TPS). These have become the foundation of the Lean process. Whenever a process is investigated, the Seven Wastes are used to drill down and identify process deficiencies. These Seven Wastes are used during the VSM exercise to differentiate between

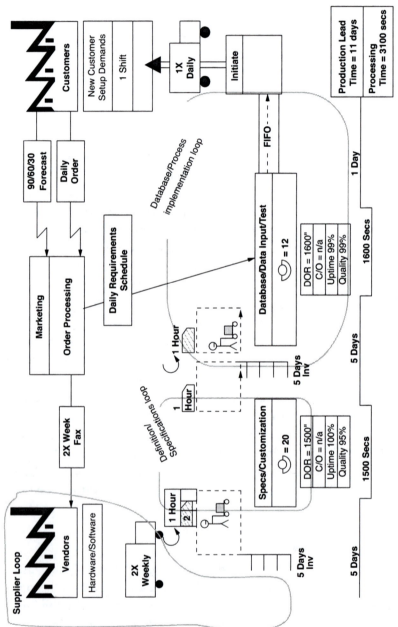

Figure 3.15 FS-VSM.

the value-added and the non-value-added activities that we are trying to eliminate. These seven categories of waste are

1. Overproduction ahead of demand
2. Waiting
3. Unnecessary transportation
4. Overprocessing due to poor product or process design
5. Inventories beyond the absolute minimum
6. Unnecessary movement by employees during the course of their work
7. Production of defective parts

Many organizations have added an eighth waste. We will include it in our discussion of the Seven Wastes. This eighth waste is

8. Underused employee abilities or creativity
   a. Overprocessing   Producing more, sooner, or faster than the "receiving" process or customer needs. Producing product for "anticipated" demand for which there are currently no orders. Producing higher-quality or more complex products than necessary.
   b. Waiting   Time delays (people waiting or queuing), process idle time, time on hand that impedes or stops work flow. Delays caused by shortages, downtime, or unnecessary (redundant) approval cycles.
   c. Unnecessary transportation   Unnecessary transportation; multi-handling; temporarily storing and moving materials, people, or information from one storage location to another. Movements that cause damage, missing items, or becoming an obstruction.
   d. Overprocessing   Unnecessary, incorrect, or redundant processing of a task. Processing higher-quality products than necessary. Adding more value to the product or service than the customer is willing to pay for.
   e. Excess inventory   Material, product, or information waiting to be processed. Producing, holding, or purchasing unnecessary inventory caused by Wastes 1 and 4, which can take up space, impact safety, and become damaged or obsolete.
   f. Unnecessary movement by employees   Excess activity (movement) or unnecessarily repeated activity. Unnecessary handling steps that could/should be automated. Poor layout (causing delays) or nonergonomic motion (causing possible injury).

   g. Production of defective parts   Rework. Products or services that do not conform to customer expectations. Correction of errors. Quality and equipment problems causing rework replacement production and scrap.

   h. Underused employee abilities or creativity   Unused or underutilized people talents. Lost time, ideas, skills, and improvements resulting from not empowering employees or tapping their creativity and talents to solve problems.

The Seven Wastes are one of the most important Lean tools. The Seven (now Eight) Wastes are brought in as appropriate to analyze the current state process. Each step in the current state process should be challenged against the Seven Wastes to assure that it is value added.

1.3.3.4.03 System Flow Chart

The VSM has given the team an excellent tool for understanding the process flow of a process. Sometimes, however, this is not sufficient for environments where the information flow is very complex. In these cases, it would be valuable to create a systems flow chart so that more of the information process can be detailed out and understood. A simple example is included in Figure 3.16.

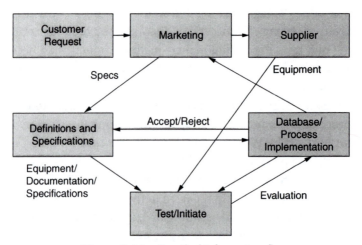

Figure 3.16 Detailed information flow.

Systems flow charts may be needed for several "systems perspectives." These include

Database perspective
User perspective
Software execution perspective

The actual flow chart can have dozens of information flow lines. It is not unusual to find out that about one-third of these lines move information forward and that the remainder are for error recovery purposes. It should become the goal of the Lean process to eliminate all the non-value-added information flow lines that exist.

1.3.3.4.04   Swimlane Chart

Similar to the systems flow chart, the swimlane chart maps out the flow of information. However, an added feature of the swimlane chart is that it shows the interactions between functional areas and the flow of information. A sample swimlane chart can be seen in Figure 3.17.

1.3.3.4.05   Spaghetti Chart

Still another mapping tool that is valuable in showing travel time for reports, materials, and/or the people involved in the process is the spaghetti chart. This is simply a floor plan of the area under consideration with lines showing the movement of people or materials for a particular process. With this diagram we can calculate travel time and travel distance, and excess travel is an enormous "waste." We should look at ways to reduce this travel time.

Figure 3.18 shows an actual example of staff travel time that was analyzed using spaghetti charting. It shows travel time and distance to get supplies, make copies, deliver reports, and so on, for a particular process. By rearranging the layout and changing the sequence of the process steps, the total travel time and distance were reduced by 75 percent.

1.3.3.4.06   5S

The objective of a 5S activity is to create an organized, safe, and productive workplace. This often requires the reorganization of the physical

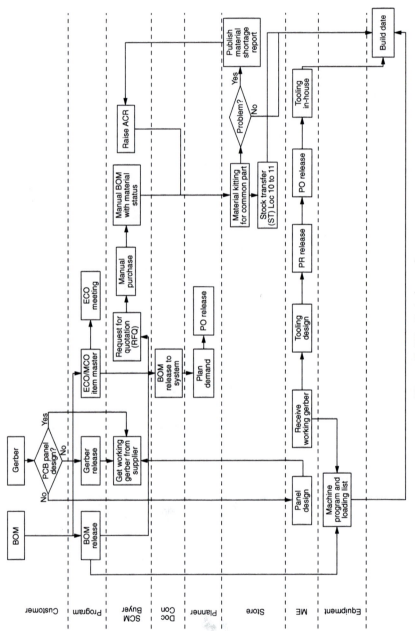

Figure 3.17 Swimlane chart.

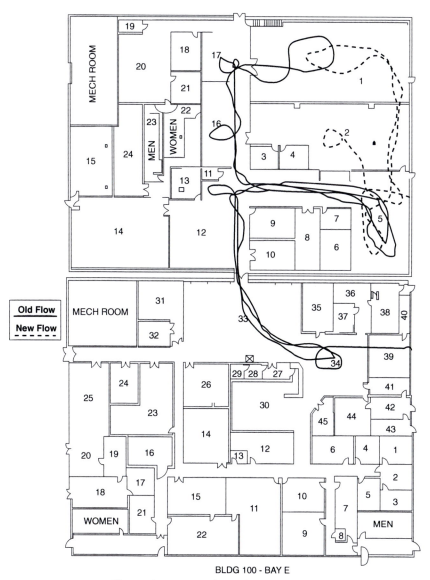

Figure 3.18   Spaghetti chart example.

workplace. It requires a change in how we move and manage our work. It often changes how current activities are performed. The 5Ss are

- Sort   Separate the needed from the unneeded items
- Set in order (Straighten)   Physically rearrange the layout and organize the work area
- Shine   Clean and remove reasons for contaminants
- Standardize   Implement procedures and signaling systems that ensure worker understanding of the process
- Sustain   Set up systems to ensure open and complete communication

SORT   With the Sort phase we start by identifying a reject area where we will place all the "tagged" items. Then we go through and question every physical item in the target area, including all equipment, inventory (piles of documents, paper, and manuals), tools, and so on. We "red tag" all tools or equipment that have not been used in the production process for more than one year, and we question all inventory items that have been in inventory for more than three months (these time periods are general and need to be adapted to the specific work environment). The slogan for this phase of 5S is: "When in Doubt, Move It Out."

SET IN ORDER (STRAIGHTEN)   The motto for the Straighten phase is: "A place for everything and everything in its place!" The process requires that we decide where we are going to keep items and then organize ourselves so that we keep these items in their appropriate locations. We need to organize not only a location, but also a methodology or a "how" we are going to keep them there. We use visual techniques for the proper identification of each work area.

SHINE   In the Shine phase we focus on preventing dirt and contamination from occurring. The objective is to create pride in the workplace, to create a safer environment, to set an environment for fewer breakdowns, and to promote a higher level of product quality.

The implementation of the Shine phase starts, as always, by analyzing the current situation. We develop a plan where we list all the areas that need cleaning.

STANDARDIZE   In the Standardize phase we focus on documenting all the 5S standards and making them visible to everyone involved in the process. We also make sure that a system exists that will maintain and monitor the work conditions to make sure that 5S standardization is continually maintained.

With the systems in place to monitor the performance of each of these items, we can develop workplace display boards that are updated regularly with photos, maps like spaghetti charts, data, and graphs. This would give everyone in the workplace a visual representation of the 5S performance improvements as they occur.

SUSTAIN   In the Sustain phase we focus on committing everyone involved in the work environment to adhere to the 5S standards. We want to establish a 5S culture of total employee involvement where 5S becomes a habit. We want 5S to become part of the corporate-wide communication plan.

A SIXTH "S"—SAFETY   In the United States, especially in the military and in military-related industries, a sixth "S" has been added, Safety, and they refer to this process as the 6S process (not to be confused with Six Sigma). In the Safety activity you would

- Look for unsafe conditions
- Look for potentially unsafe acts
- Look for difficult tasks (are they ergonomic?)
- Try the jobs yourself (where could you get hurt?)
- List the opportunities
- Resolve them

The 5S process begins with a scan of the current workplace. The objective of the scan is to document current operating conditions. We clearly define the target area. Then we try to define the purpose and function of the targeted area. We use our maps (VSM, spaghetti chart, and system flow chart) to show the physical flow between equipment and people. We look for problem areas, like inconsistent or intermittent flow, quality failures, or bottlenecks. We record current problems on a checklist. We photograph and document these problem areas and put them on a display board so the entire team can review and evaluate these areas.

1.3.3.4.07   Cellular Work Design

Traditionally, organizations are laid out by functional areas or departments. Work flows through the organization by being moved from one department to another. In Figure 3.19 we show how most organizations believe the flow of materials or people occurs in their organizations. After the VSM exercise they realize that the flow is more complex then they

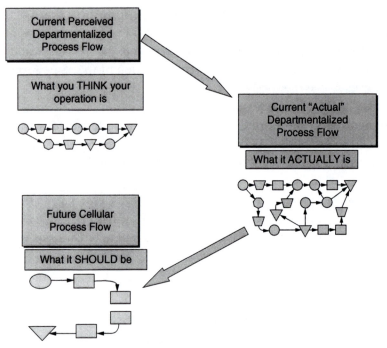

Figure 3.19 Traditional versus cellular design.

at first thought it was. Bringing cellular design concepts into the Lean exercise teaches the attendees that the future state of the organization needs to organize work by the way the product flows through the process. Departments are arbitrary barriers to smooth flow. Product flow organized in the sequence in which production occurs is more efficient and eliminates waste.

1.3.3.4.08   JIT—Just-in-Time

Just-in-Time (JIT) systems bring in only what is needed, when it is needed, and in the amount it is needed. JIT does not believe in stock piling anything. It attempts to minimize materials, equipment, labor, and space. Ideally, JIT wants employees to only have the one piece they are working on, nothing else. JIT utilizes the pull system, where nothing is introduced into the process until it is requested. Materials are "pulled" from their source and never "pushed" into the process based on a schedule.

A typical JIT company will utilize small, inexpensive machines in the process, not large, multifunction ones. It will ideally have one-piece project flow, not batches of projects. It will have standing operators that move around between functional steps in the process. It will continually look for waste reduction opportunities in the process.

It would require several books to cover JIT adequately. There are literally hundreds of books available on this subject. The author will leave it to the reader to choose from these sources for further in-depth reading and understanding.

1.3.3.4.09   Agile

The term "agile" is used in reference to everything from manufacturing to project management to software development. Unfortunately, it means different things depending on what is being referenced. For our purposes we will refer to agile manufacturing or agile processes. In this case, agile means flexibility. Specifically, an agile process is one that can easily be disrupted. In the case of manufacturing, an agile process is one where the product being manufactured can easily be reconfigured to produce an entirely different product.

A successful agile process would be one in which there is no committed inventory. The work-in-process inventory can easily be reconfigured for a different end product. Agile is also where cycle times are extremely short, thereby allowing rapid flexibility in the process. Agile requires a combination of Lean, JIT, and a pull production environment.

1.3.3.4.10   Poka-Yoke

Poka-yoke is a Japanese term that means "mistake proofing" or "error proofing." It is a method of redesigning processes to prevent errors and is one of the foundational quality improvement tools in any Lean process. Some real-life examples of poka-yoke tools include the following:

- Computer disks that can only be inserted in the drive one way
- Automatic seat belts
- Irons that automatically shut off
- Plugs that will only go into the socket one way

Poka-yoke strives to make inspections obsolete. With poka-yoke you would redesign the process so that it will validate and eliminate quality errors. There is only one way to do the process, forcing you to do it the right way.

1.3.3.4.11 PQ Analysis

The product/quantity (PQ) analysis is the tool used to identify the products (P) and the quantities (Q) that will be going through the production cell. We are attempting to forecast the volume of activity that the cell will need to manage.

A bar graph is made showing the individual contribution and the cumulative contribution of all the parts going through a specific cell. This graph is later used to facilitate the design and development of the cell.

1.3.3.4.12 Total Product (Productive, Preventative, or Production) Maintenance

Total product maintenance (TPM) (this acronym is also used for total productive maintenance, total preventative maintenance, or total production maintenance) is a system focused on helping to attain and maintain competitiveness in quality, cost, and delivery. TPM focuses on strategies for creating employee ownership of the process and the equipment, thereby generating in employees an urgency to eliminate system waste. TPM is utilized to alleviate system wastes associated with equipment (improving the effectiveness and longevity of machines). It is an attempt to eliminate minor work stoppages and focuses on zero defects.

1.3.3.4.13 Visual Workplace

Visual workplace is a concept pulled out of 5S that focuses on "a place for everything and everything in its place." It stresses labeling storage locations and clearly demarking them so there is no confusion about where anything belongs. It utilizes andon lights in the workplace so that work stoppages and work problems can easily be identified. It uses display boards to increase the visibility of work performance, including quality and productivity. And it focuses on pictographic training tools so that employees can easily see how the work is to be done.

1.3.3.4.14 DFM—Design for Manufacturability

DFM (also referred to as DFA—design for assembly) is a tool used to design error-proofing and improve ease of assembly into product engineering. It has always been a component of concurrent engineering and isn't exclusively a Lean tool. The focus is that a design isn't beautiful if it isn't practical. DFM requires collaboration between design, manufacturing, and the customer, and the end design is an integration of each of these perspectives.

Engineering no longer dominates the design process. Issues like "tolerancing" take on a manufacturing perspective, rather than being some arbitrary "engineering standard."

Another important element of DFM is that the product is designed in such a way that it can't be assembled incorrectly. In the end, DFM improves overall product quality, reduces production mistakes, and reduces fabrication costs.

1.3.3.4.15   SMED—Single Minute Exchange of Die, Quick Changeover

SMED is focused on mastering the ability for quick changeover. The goal is to minimize non-value-added time, which is the downtime associated with the end of one value-added productive activity and the start of the next productive activity. Lean recommends doing a VSM that maps out the steps involved in these activities and focuses on eliminating or minimizing as many of these steps as possible.

1.3.3.4.16   Kanban

The kanban is a card—specifically a tracking card—used to monitor the flow of materials through a JIT pull environment. Materials cannot exist in the system unless they have a kanban card attached to them. Through this process, the kanban card becomes an inventory control mechanism. The size of the kanban (how many parts are attached to it) is carefully calculated utilizing EOQ (economic order quantity) principles.

There are specific inventory staging areas in the production process, and each staging area can have at most one kanban card in it. A kanban cannot move forward in the process unless the next staging area has been emptied. Otherwise, it must wait at the previous station until the next slot becomes available. In this way, production always knows exactly how much inventory is in the process and where it is located.

1.3.3.4.17   Jidoka

Jidoka means "build in quality at the source." This is one of the fundamental principles that brought world attention to TPS because of its extremely high levels of quality. Quality is not the focus of inspections and control points; quality occurs as an integral part of the production process. After work has left the station, it is too late. Quality has to be validated before the item leaves the station. Defects are never passed up the production stream.

In jidoka we utilize a variety of root cause analysis tools. We also use mistake-proofing, which attempts to create a production environment where a process can only occur one way—the correct way. Visual management is also utilized, which makes "doing the wrong thing" visually obvious.

The basic principle behind jidoka is that the person who does the work is responsible for the quality of the work. If the person doing the work can't fix the mistake, then it is his or her responsibility to stop the work flow until the problem can be properly addressed.

1.3.3.4.18 Standard Work

Standard work focuses on developing a standardized methodology for each process so that everyone who performs that activity will perform it in the same way. Standard work is used to identify the best known way to complete a task and then teaches everyone that "standard" method. Standard work ensures that the same work will take the same amount of resources to achieve the same results every time. If work processes are not standardized across the organization, it is impossible to effectively forecast resource utilization.

Standard work is a foundational tool of continuous improvement. Once discovered and implemented, the team should standardize on this improved process across all team members so everyone will benefit from using this best practice.

The standard work for every process step will be unique. Here are some general guidelines that facilitate a standardized process:

1. Involve staff from all shifts, all work centers, and all locations that do the same type of work.
2. Let the process workers come together to define the work and gain consensus.
3. Keep it simple. Unnecessary complexity adds opportunities for failure.
4. Document the standard work and train from the documentation.

1.3.3.4.19 Brainstorming

Brainstorming is a tool whereby a team generates a large number of ideas in a short amount of time. It encourages team members to generate ideas in an open setting. By setting rules about idea generation—for example, that there are no bad ideas—the team is freed from fear of judgment (retribution and attribution are eliminated). Doing brainstorming in a team setting lets team members build on, and be inspired by, each other's ideas. The technique is especially useful in finding creative new approaches to difficult problems.

Brainstorming occurs in a group or team setting, ideally between 6 and 12 members. The team should throw out any idea they can think of without being considered outlandish.

1.3.3.4.20   Fishbone Charting

A fishbone diagram (or chart) is a method for organizing and analyzing all contributors to a problem. Fishbone diagrams literally look like a fish skeleton, with six ribs branching off a central spine. A problem statement serves as the fish head, with the six ribs representing the six broad categories of causal factors: Manpower, Machine, Method, Material, Measurement, and Environment. These six factors are not rigid and may need modification in order to fit a specific application. An example of a Fishbone diagram can be seen in Figure 3.20.

A fishbone diagram is most effectively completed by using a cross-functional team of process and customer experts who are the most familiar with the problem. Ideally, the team size is from 6 to 12 members. The procedure is as follows:

- Draw the skeleton of the fish to be filled in.
- Label the fish skeleton. The head contains the problem statement, and the ribs are labeled Machine, Material, Measurement, Method, Manpower, and Environment.
- Have the team brainstorm ideas that could be contributing to the problem.
- Write each idea on a separate sticky note.
- Place each sticky note on the rib that most closely describes the category the idea relates to.
- After a free-flowing stream of ideas dries up, have the team concentrate on each rib, one at a time, to try and squeeze a few remaining ideas out of the team.

1.3.3.4.21   Eight-Step Problem Solving

Eight-step problem solving works in conjunction with the A3 reporting methodology. The format outlined here is not the only A3 format available. There are many others, and you can find them in a variety of Lean books. However,

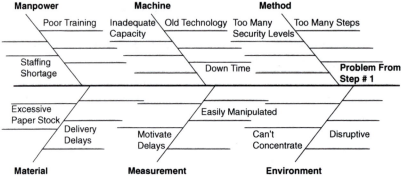

Figure 3.20 Fishbone chart.

the author has experienced a great deal of success using the example of the tool presented here, which has nine steps, and is used for both opportunity analysis (the first seven steps) and later for governance reporting (steps 8 and 9).

The eight-step (nine-step in this case) report looks like the example in Figure 3.21. This report format is used for opportunity analysis, project presentation, project justification, and project performance review. When

Team Members:	9-Step Opportunity (Problem) Analysis Tool	Approval Information/ Signatures
1. Clarify and validate the problem	5. Determine root cause	7. Execute improvement tasks
2. Perform a purpose expansion on the problem		
3. Break down the problem/identify performance gaps	6. Develop improvement task list	8. Confirm results
4. Set improvement targets		9. Standardize successful processes

Figure 3.21 9-Step A3.

you use the same format consistently, the entire organization (management and users) become familiar with the format and you don't waste time explaining each box. Everyone knows what they are looking for and they go right to the box that interests them.

At this point we will go through each of the boxes in the A3. We need to take a "big picture" look at the nine-step A3 tool and see how it works. Figure 3.22 offers a brief description of the purpose of each of the boxes.

"Team Members" Box   When analyzing a problem or opportunity it is important to include all the relevant subject matter experts (SMEs). The "Team Members" box lists the individuals that were involved in the analysis work that went into the creation of the nine-step A3 document.

"Approval Information/Signatures" Box   Every improvement project must have a Champion(s). This is the individual who controls the purse strings. This Champion must be willing to sign his or her name on the dotted line approving this effort. If you don't have a Champion, then don't waste your time on the project.

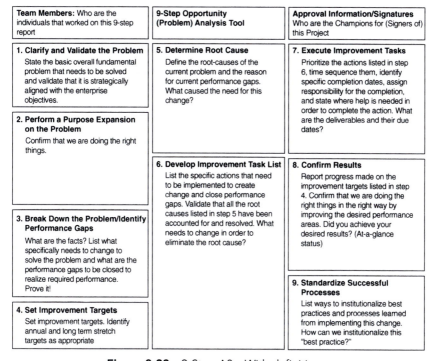

Figure 3.22   9-Step A3—With definitions.

"Clarify & Validate the Problem" Box   This box defines the opportunity or improvement project that you are about to engage in. In this box we state the fundamental opportunity that we want to work on and define it sufficiently so that anyone looking at this A3 will know what we're talking about.

"Perform a Purpose Expansion on the Problem" Box   This step is normally not a part of the eight-step process. However, the author has had enormous success adding this step and has included it in this discussion. The use of this step is most easily demonstrated with an example. Mitsubishi had all their employees log all their activities, every day, for one week. After this exercise, they had each type of activity analyzed by adding three columns to the right of each activity. The columns were as follows:

First Column: What was the purpose of this activity?
Second Column: What was the purpose of the purpose listed in the previous column?
Third Column: Does the purpose in the previous column achieve either of the following:

1. Increase customer satisfaction?
2. Increase the quality or performance of the product or service we perform?

If the answer to the question in this last column was "No," they were told to quit performing that activity. Approximately one-third of the activities of the enterprise were eliminated.

"Break Down the Problem/Identify Performance Gaps" Box   Box 2 of the nine-step A3 report had us look upward in the organization. Box 3 and onward has us looking down deeper into the problem or opportunity that we are exploring. The objective of box 3 is to answer the questions "What makes this problem a problem?" and "Can you prove that it really is a problem?"

"Set Improvement Targets" Box   The key questions that we need to answer in this box are "What is the goal?" and "What do you want to achieve by making this change?" In this box we utilize some of the KPIs or metrics that we identified in box 3 and set targets that we hope to achieve.

"Determine Root Cause" Box   In this box we focus on defining the root causes of the current problem and the reasons for the current performance gaps. We try to answer the question "What caused the need for this change?" We look to find out what is the immediate thing that is happening or not happening that is creating the problem.

"Develop Improvement Task List" Box   Up to now we have been focused on making sure we are doing the right thing. Are we fixing the right problem? Are we focused on the correct opportunity? Starting with Box 6 we shift our focus away from "doing the right thing" and look at "doing things right." But we did not do boxes 1 through 5 in a vacuum. As we worked our way through the last five steps we identified improvement opportunities. Particularly in Step 4 when we developed the CS-VSM, we noticed a large number of things that did not make sense. Now, in box 6, we record all these "opportunities for improvement" and identify tasks showing how we should improve this process.

"Execute Improvement Tasks" Box   This box answers the question "Why aren't we there yet?" It takes the tasks that were identified in box 6 and performs the following activities:

1. Prioritizes them based on their impact on the organization and effort to install
2. Assigns a due date
3. Assigns a department or person of primary and secondary responsibility (you can't commit someone who isn't present—if they're not part of the team, they need to be contacted and be brought into the team)
4. Identifies areas where specific help is needed that goes beyond the scope of this team (this will become the responsibility of the team Champion, who was listed in the upper-right corner of the nine-step A3 document)

"Confirm Results" Box   The purpose of this box is to prove that we're working on the right things; that we're actually having the impact that we discussed in boxes 1 through 5. We start with the improvement targets that were set in box 4, and we monitor and measure our performance to those targets.

"Standardize Successful Processes" Box   This box is about communication. It focuses on sharing the lessons that we learned during this nine-step A3 exercise. An important part of any Lean effort is to make sure that improvement efforts are not occurring in isolation. The lessons learned should be standardized and shared.

1.3.3.4.22 SWOT—Strengths, Weaknesses, Opportunities, Threats

SWOT is an excellent tool for structuring a brainstorming exercise. SWOT analysis provides a strategic perspective of the organization in its current state. It is used to create a list of political, environmental, technical, managerial, or programmatic issues in an orderly format during strategic planning sessions. SWOT analysis provides input to strategic planning and forces discussion and assessment of internal (Strengths, Weaknesses) and external (Opportunities, Threats) issues that are affecting the organization. It also forces the team to prioritize the issues within each category and drives toward consensus.

1.3.3.4.23 Voice of the Customer

Voice of the customer (VOC) is a tool that gives the user the customer's perspective as to what is important and what is not. The customer defines what is value added and what is waste. The first step in improving a process is to understand the customer's needs. Gathering the VOC has two parts: 1) identifying the customer and 2) documenting the VOC.

To better serve your customer, you must determine your customer's needs. For example, what are your customer's critical to quality (CTQ) issues? CTQs are those basic stated requirements that "must" be met in order to satisfy the customer. Your customer may also have unstated or implied needs that they take for granted and do not feel they need to specify.

1.3.3.4.24 Gemba Walk—Go and See

The gemba walk is a critical tool when developing the eight-step A3 or when doing a VSM. After the goals of the process improvement activity have been established and the SIPOC has been performed, defining the scope of the improvement effort, it now becomes important to physically go out and observe the process that needs to be improved. The team is ready to observe the process and to gather data on it. This requires a detailed walk through every step of the value stream. The team takes detailed notes on the flow of the data, paper, materials, information, and people as they move through this process.

After the interview process the team is now ready to create the VSM. A table is created on the VSM for each step in the process, which has all the relevant information that will be needed to analyze the value stream.

1.3.3.4.25   Gap Analysis

Gap analysis is where we look to the future (FS-VSM) and compare it with the current state (CS-VSM). Then we ask the question "What is it going to take to get us from the CS to the FS?" These steps become the gaps. After listing the gaps we identify their complexity and feasibility. We study the cost of correcting the gap and we prioritize the gaps. From this information we can then develop an implementation plan.

1.3.3.4.26   Five Whys

The Five Whys is a simple method to help problem-solving teams drill down to the "true" root causes. To prevent problem recurrence, teams must address the root causes of the problems rather than just the symptoms. Teams should not take "five" as an absolute. They may need to drill deeper than five. Rarely do they identify true root causes in fewer than five steps.

A strength of the Five Whys tool is its simplicity. The tool starts with a simple problem statement and asks "Why did the problem occur?" When an explanation is reached, the team asks "why" again . . . five times, each iteration asking "why" about the previous explanation.

1.3.3.4.27   SIPOC or COPIS

In SIPOC we define the process under study by looking at and analyzing all the pieces (Supplier—Input—Process—Output—Customer). This is used to define the process value stream and identify opportunities for waste elimination. In Six Sigma we find the term COPIS, which is SIPOC in reverse. Each letter means the same thing.

When working on a Lean improvement process, we need to take a "systems" perspective. What is the system that we are analyzing? Where does the process start, and where does it end? Once we have drawn a circle around the system under consideration, we can then look at the internal process and the external influences to this system. SIPOC is a tool to help the Lean team work through this "systems" thought process.

1.3.3.4.28   Pull Signaling

A pull system is one where production does not move forward until it is pulled forward by a demand somewhere down the supply chain. JIT and kanban are both considered pull-based systems. The antithesis of pull is push, where production is triggered based on a schedule. If the schedule says it's time to produce something, then we produce it. In contrast, in a pull environment, production does not begin until there is a customer order, which will "pull" the need to produce through the production process.

Some characteristic benefits offered by a pull environment over a push environment include reduced inventory, reduced cycle time, less facilities space, and improved capacity utilization, just to name a few.

1.3.3.4.29 Affinity Diagrams

Affinity diagrams facilitate a simple and flexible way to organize ideas and information. If the team is unfamiliar with a problem area that they are working on or feels overwhelmed by a large volume of data or ideas, affinity diagrams can bring order to the chaos. Affinity diagrams simply lump ideas together. By grouping and regrouping ideas and data points, teams can begin to organize patterns of thought from incoherent information.

Affinity diagrams are often used in conjunction with the brainstorming. The large volume of loosely related ideas generated by brainstorming can be organized using affinity diagramming. Fishbone diagrams are a specific form of affinity diagrams.

1.3.4 Shingo Prize

We are nearing the end of our definitions of quality systems and tools. Table 3.10 shows the remaining quality topics that we will need to define under TPS.

Before we leave our discussion of Lean tools, we need to discuss the "Lean standard" for excellence: the Shingo Prize for Excellence in Lean

Table 3.10 Quality Methodology

1	TPS—Toyota Production Systems
1.3	Lean
1.3.4	Shingo Prize
1.4	Six Sigma
1.4.1	DMAIC
1.4.2	SQC—Statistical Quality Control
1.4.3	SPC—Statistical Process Control
1.4.4	Queuing Theory
1.4.5	DFSS
1.5	QFD—Quality Functional Deployment
1.6	Concept Management

Manufacturing. The Shingo Prize has developed a model that defines excellence. However, in the spirit of continuous improvement (CI), as the Lean methodology develops, the Shingo model also continues to evolve. The current state of the model is shown in Figure 3.23.

The Shingo Transformation Model evaluates Lean performance in the following "dimensions" (see Utah State University, College of Business, *The*

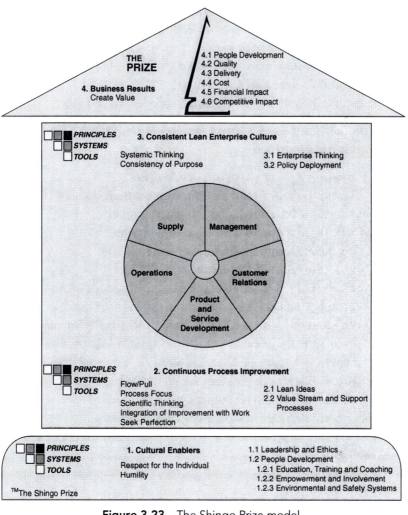

Figure 3.23   The Shingo Prize model.

Shingo Prize for Operational Excellence—Model—Application Guidelines,
3rd Edition 2008, www.shingoprize.org) (also discussed on page 55):

> Let's look at the power that this collection of tools is capable of.
> Recently, the author was the facilitator for a Lean process that
> involved engineering and design, supply chain logistics, contract-
> ing, IT in a systems redesign, a major construction and remodel
> of the facility, and the automation of an extensive documentation
> process. Some of the results of this one-year activity included

▲ A 75 percent reduction in travel time
▲ A reduction in queue levels from 200 jobs to 14 jobs, with a goal of
moving to below 10 jobs
▲ A reduction in flow days from 141 days down to 81 days before the
remodel and 45 days after the remodel
▲ A reduction in "jobs on hold" from 14 jobs down to 1 job

Additional examples of related Lean process improvements where the
author was involved include

▲ A 40+ percent decrease in process cycle times
▲ A 64 percent lead-time reduction for forecast development
▲ A 75 percent lead-time reduction in scheduling kits
▲ A 90 percent reduction in non-value-added time
▲ A 93 percent reduction in front office research time
▲ A 93 percent reduction in order processing time
▲ A 70 percent reduction in lead time from the logistics carrier
▲ A 99 percent reduction in equipment maintenance flow days
▲ A 99 percent increase in medical records processing efficiencies
▲ A 90+ percent accuracy in data input quality (formerly less than
20 percent)
▲ A 98 percent increase in accuracy for digital orders
▲ A 76 percent decrease in screening cycle time
▲ A 100 percent decrease in laboratory labeling errors
▲ A 100 percent improvement in patient safety from reduced labeling
errors
▲ A 200 percent increase in operating room utilization due to improved,
accurate scheduling

And the success stories go on and on in all areas of industry.

1.4 Six Sigma

The TPS used the term Six Sigma $(6\sigma)^2$ as a performance measure and goal. It is now utilized to accurately measure quality. But it has become more than that. It has become a CPI practice and quality improvement program of its own. By using Six Sigma we can set process goals in parts per million (PPM) in all areas of the production process. Since its origin, Six Sigma has evolved into a methodology for improving business efficiency and effectiveness by focusing on productivity, cost reduction, and enhanced quality.

Six Sigma has its roots in the TQM efforts of Joseph Juran and W. Edwards Deming. Their programs for zero defects and total quality management in Japan lad to Motorola adopting the Six Sigma philosophy. Motorola was able to achieve a 200-fold improvement in production quality and saved a reported $2.2 billion using this tool. Today Six Sigma has evolved to become a management methodology that utilizes measures as a foundational tool for business process reengineering.

The name comes from the statistical use of the sigma (σ) symbol, which denotes standard deviations. Six identifies the number of standard deviations around the mean. Hence, in Six Sigma we are saying that you have to go out beyond six standard deviations around the mean before you find failure. With a high enough number of sigmas (beyond six), you would approach the point of "zero defects." The sigma levels step changes—for example, moving from Three Sigma (93 percent accuracy) to Four Sigma, requires quantum leaps of improvement. A move from Three Sigma to Four Sigma is an 11-fold improvement. At the Six Sigma level, the product failures (number of parts beyond the allowable limits) would be 3.4 parts per million. This equates to a 99.9997 percent accuracy. In today's world, 98 percent or 99 percent accuracy is considered excellent. However, Six Sigma has now become the universally recognized standard of quality.

Some guiding principles of the Six Sigma CPI program are

1. If you want something to happen, you better measure it. Unfortunately, that also means that if you measure the wrong things, you'll get the wrong results. For example, measuring throughput may speed up production, but at the cost of quality. Measuring quality may increase

---

[2] Some of this material was developed by the author for an article in the *Encyclopedia of Management* (Marilyn M. Helms (editor), "Six Sigma and SPC," *Encyclopedia of Management*, Gale Group Publishers, 2006. pp. 816–821).

quality, but decrease customer service. So one of the toughest challenges in Six Sigma measurement, and one of the biggest reasons for failed Six Sigma projects, is identifying a measurement system that will trigger the correct collection of responses.

2. All the measures should be openly visible. Openly displaying all measures on charts and graphs is a primary motivator toward the correct response.

3. The change curve applies. When change happens, performance will initially go down before it recovers and goes back up.

4. CPI requires cultural change or change readiness. If the organization is not primed for change, then an environment for change must be instilled prior to starting Six Sigma, or the project is doomed to failure.

Six Sigma concentrates on measuring and improving those outputs that are critical to the customer. The tools to accomplish this include a range of statistical methodologies, which are focused on continuous improvement using statistical thinking. This paradigm includes the following principles:

▲ Everything is a process.
▲ All processes have variations that are inherent within them.
▲ Data analysis is a key tool in understanding the variations in the process and in identifying improvement opportunities.

The Six Sigma CPI process incorporates a problem solving and process optimization methodology. Six Sigma creates a leadership vision utilizing a set of metrics and goals to improve business results by using a systematic, five-phased problem-solving methodology. Two common problem-solving project management methodologies are commonly associated with Six Sigma. The first is DMAIC (Define, Measure, Analyze, Improve, Control), and the second is DMADV (Define, Measure, Analyze, Design, Verify). Later we will discuss the most common, DMAIC.

Statistical process control (SPC) is a foundational statistical analysis tool for Six Sigma. Both SPC and its companion, statistical quality control (SQC), are tools utilized by a Six Sigma improvement process to provide productivity and quality information about a production process. The SPC and SQC processes collect data and report results as the process is occurring so that immediate action can be taken.

Some excellent and highly recommend websites include www .onesixsigma.com and www.qualitydigest.com, which include several informative articles.

1.4.1 DMAIC

As mentioned, the two common problem-solving project management methodologies commonly associated with Six Sigma are DMAIC (Define, Measure, Analyze, Improve, Control) and DMADV (Define, Measure, Analyze, Design, Verify). This book will discuss the most common, DMAIC.

Tables 3.6 and 3.8 at the start of this chapter compared the Six Sigma DMAIC with the Lean PDCA and TOC loops. From these tables, we see that all three are effectively the same.

Six Sigma is a measurement-based strategy that focuses on reducing variations. It holds that every process should be repeatedly evaluated and significantly improved, with a focus on time required, resources, quality, cost, and so on. The philosophy prepares employees with the best available problem-solving tools and methodologies using the five-phased DMAIC process. Reviewing each of the DMAIC steps in more detail, we have the following:

▲ **Define**   At the first stages of the process we look for and identify poorly performing areas of a company. We then target the projects with the best return and develop articulated problem and objective statements that have a positive financial impact on the company. We "define" the opportunity from both the organization and the customer perspectives.

▲ **Measure**   At this stage we are trying to tie down the process under consideration. Where does it start and end? What should we be measuring to identify the deviation? What data characteristics are repeatable and identifiable? What is the capability of the process? We develop a baseline for the targeted area and implement an appropriate measurement system.

▲ **Analyze**   Having identified the "who" and "what" of this problem, we now target the where, when, and why of the defects in the process. We use appropriate statistical analysis tools, scatter plots, SPC and SQC, input/output matrixes, hypothesis testing, and so on, and attempt to accurately understand what is happening in the process.

▲ **Improve**   At this point we should have identified the critical factors that are causing failure in the process. And, through the use of experiments, we can systematically design a corrective process that should generate the desired level of improvement.

▲ **Control**   In the control phase we implement process control tools that can manage and monitor the process on an on-going basis. The DMAIC process is now in full operation, but it does not stop here. We need to develop a control plan to continuously monitor the process.

1.4.2   SQC—Statistical Quality Control

SQC utilizes control charts, which are a key tool for Six Sigma. Control charts are also referred to as statistical process control (SPC)[3], which is a foundational statistical analysis tool for Six Sigma. The original objective of SPC and SQC is to provide productivity and quality information about a production process in real time. The focus is on process control and continuous improvement. The operators become their own inspectors and control their own processes.

Control charts visually organize time series data collected as observations of a process. The visual display of quantitative data makes it easier to observe how a process is behaving. It differentiates between variation due to "noise" (also called common cause variation) and variation that signals a significant change in the process (also called special cause variation).

In a control chart, the vertical axis usually represents some quantitative measure of the process output as sampled at a fixed interval. The chart typically has three equally spaced horizontal lines running across it. The center line is the expected average of the measurements, the upper line is the upper control limit (UCL), and the lower line is the lower control limit (LCL). Data points that fall above or below the control limits indicate a change has occurred in the process and an adjustment should be made. As long as the data points fall between the UCL and LCL, the process is behaving normally and no action should be taken.

The next step in the SPC process is to establish a set of control variables, which includes an average (X) and a range (R). These can be established by going to the drawings and reviewing the initial part specifications using the expected value as X and the tolerance range as R. Or, these variables can be established using historical values and calculating the historical average (X) and range (R) for the data.

---

[3] Some of this material was developed by the author for an article in the *Encyclopedia of Management*.

Having established an X and R value, we can calculate a UCL and an LCL.

$$UCL = X + R$$
$$LCL = X - R$$

From these values, a pair of control charts is created. These charts are used to plot the SPC data as it occurs and provide a visual tool to monitor the process. Figure 3.24 is an example of the X-bar SPC chart that monitors a process. For this chart we will use X = 1.23 and R = 0.45.

From Figure 3.24 we can see how the measurement data is recorded on the chart at the time each measurement occurs. The objectives behind this data collection process are several. One is to catch outliers in the data (anything above the UCL or below the LCL). These outliers are quality failures that immediately stop the process. Another purpose for the measures is to identify trends. For example, data points 1 through 5 indicate a strong trend to failure approaching the LCL. Corrective action should be taken immediately to avoid the possibility of producing bad parts. Another objective can be seen in data points 7 through 13, which indicates that perhaps our LCL and UCL are too far and need to be brought in tighter, thereby giving us a higher level of performance and a higher level of quality.

Another methodology for applying SPC processes is by collecting data, not on every event, but on a random sampling of the events. This occurs

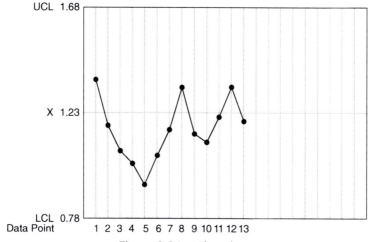

Figure 3.24   X-bar chart.

when there is a large volume of activity and the time required to measure each event is too burdensome. A statistical sample is taken, and from that sample the average of the sample data (X_1) and the range of that sample $(R_1 =$ highest minus lowest measure) are calculated. Using the statistical random sample, a range chart (R-chart) would also need to be created.

With the X-bar and R charts, we can now create summarized reports, like the histograms and frequency distributions that were discussed earlier. This allows a long-term, summarized perspective of the process, rather than the chronological timeline that the X-bar and R charts offer.

Once the control chart is set up, the person closest to the work should be placed in charge of taking the samples and entering the data in the chart. UCL and LCL values can be determined mathematically based on sample size and standard deviation of the data points already collected. UCL and LCL values can also be identified as customer specifications.

1.4.3 SPC—Statistical Process Control

The SPC[4] process collects data and reports results as the process is occurring so that immediate action can be taken. When appropriately applied, SPC can virtually eliminate the production of defective parts. It makes the cause of the failure visible. Since an operator is able to immediately recognize that a failure is occurring, he would be able to react to that failure, observe its cause, and then take corrective action. As Peter Drucker emphasizes, the "operators become the 'owners' of not just the process, but also the parts they produce."

Because of its success, SPC has found application in other industries, including service industries, transportation industries, and delivery services, and can even be found in fast food and baggage handling. For example, on-time delivery performance can be monitored on an SPC chart.

Several tools are available for the display of SPC data. These include the following:

1. Graphs and charts are used to display trends or to summarize the data. These tend to be bar or line graphs that report on a specific parameter of performance.

---

[4] Some of this material was developed by the author for an article in the *Encyclopedia of Management*.

2. Check sheets or tally sheets are used to take the raw data and reorganize it into specific categories that are being observed.
3. Histograms or frequency distribution charts are used to translate raw data into a pictorial display showing the performance of specific quality characteristics.
4. Pareto principles are used to prioritize the contributory effect of specific quality problems. This tool assists in identifying which problems have the largest impact on a specific quality problem under study.
5. Brainstorming is used to generate ideas by taking advantage of the synergistic power of a team of people.
6. Ishikawa diagrams (fishbone charts) are used to create problem and solution visibility by grouping problem causes into branches. Often this is referred to as a cause-and-effect diagram. Using this tool in conjunction with the PDCA process helps to narrow down the root cause.
7. Control charts are used to validate that the variation of measurement of a specific parameter is kept within a set of control limits.

1.4.4   Queuing Theory

Queuing theory is a mathematical operations research tool used for studying waiting lines, or queues. The theory facilitates the analysis of several interacting processes, including arrival to the queue, waiting, and being served. It also facilitates the calculation of metrics, such as the average waiting time in the queue, the expected number of individuals waiting or receiving service, and the probability of certain system "states," such as empty or full.

Queuing theory is often used when making business decisions about the service resources needed. It is applicable in business, industry, healthcare, public service, and engineering. Applications are frequently encountered in customer service, logistics, and communications. Applying simulation and modeling to queuing theory allows organizations to select the optimal service structure that will maximize customer satisfaction.

1.4.5   DFSS

Design for Six Sigma (DFSS) is an outgrowth of Six Sigma, and is a newer variation of TPS. Adherents to DFSS like to consider it a separate and emerging business-process management methodology. Six Sigma is extremely tool-based whereas DFSS is closer to Lean. It focuses on the needs of customers.

Where Six Sigma uses DMAIC, the DFSS loop focuses on DMADV. DMAIC usually occurs after the initial system or product design and development have been largely completed. It normally focuses on solving existing manufacturing or service process problems and removing variation, which causes defects. DMADV strives to generate a new or improved process similar to Lean. DFSS aims to create a process that integrates the process improvements of Lean with the efficiencies of Six Sigma.

DFSS uses advanced VOC techniques and systems engineering techniques to avoid process problems. It yields products and services with greater customer satisfaction. DFSS techniques also include tools and processes to predict, model, and simulate the product delivery system as well as the system life cycle.

1.5 QFD—Quality Functional Deployment

QFD is the implementation of a continuous improvement process focusing on the customer. It was developed at Mitsubishi's Kobe Shipyards and focuses on directing the efforts of all functional areas on a common goal.

QFD systematizes the product's attributes in a matrix diagram called a "house of quality" and highlights which of these attributes is the most important to a customer. This helps the teams throughout the organization focus on their goal (customer satisfaction) whenever they are making change decisions, like product development and improvement decisions.

QFD focuses on

1. The customer
2. Systemizing the customer satisfaction process by developing a matrix for defining
   a. Customer quality
   b. Product characteristics
   c. Process characteristics
   d. Process control characteristics
3. Empowered teaming
4. Extensive front-end analysis, which involves 14 steps in defining the "house of quality":
   a. Create and communicate a project objective
   b. Establish the scope of the project
   c. Obtain customer requirements

d. Categorize customer requirements
e. Prioritize customer requirements
f. Assess competitive position
g. Develop design requirements
h. Determine relationship between design requirements and customer requirements
i. Assess competitive position in terms of design requirements
j. Calculate importance of design requirements
k. Establish target values for design
l. Determine correlations between design requirements
m. Finalize target values for design
n. Develop the other matrices

Implementing and using QFD is not an easy process. A great deal of commitment throughout the company is required for the process to be successful. The results of effective implementation are well worth the effort. Reduced product development time, increased flexibility, increased customer satisfaction, and lower start-up costs are just a few of the benefits that can be expected through the use of QFD.

GREGG D. STOCKER

1.6   Concept Management

Breakthrough thinking and concept management (CM) are tools that developed sequentially. Breakthrough thinking is a revolutionary tool for brainstorming or idea creation developed by Shozo Habino. Concept management, which has Japanese roots, integrates TPS tools like TQM and World Class Management (WCM) with breakthrough thinking. All of these tools can be utilized as a CPI change model that focuses on (1) innovation and creativity, and (2) making sure we are "doing the right things." These tools focus on asking "Why are we doing this?" using the "purpose expansions" before asking "Are we doing this right?" which tends to be the focus of the "root cause analysis."

The breakthrough thinking utilized by CM moves away from the slowness and costliness of traditional root cause analysis commonly used in the United States and Europe. WCM offers the formal structure around which the ideas are turned into goals and a measurement/motivation system.

TQM is the process for team-based ideas and change implementation. Thus, CM is an idea generation and implementation process used by companies like Toyota and Sony that breaks us out of the traditional, analytical thinking common to companies such as the Ford Motor Company, which uses the TOPS program, or the Russian TRIZ program. Instead, it focuses on forming a purpose hierarchy through a series of steps.

World Class Management is broad in its application, and numerous publications discuss the subject in detail (see Plenert's book *The eManager* [Blackhall Publishing, 2001] or his book *Making Innovation Happen; Concept Management Through Integration* [CRC Press, 1997]). World Class Management is not a system or a procedure; it is a culture. It is a continually molding process of change and improvement. It is a competitive strategy for success.

In the United States, TQM has fallen into disfavor because of its analytical approach to change. The analysis process is deemed too slow to be competitive, but that is primarily because TQM utilized root cause analysis. With breakthrough thinking we can revisit our use of TQM.

Concept management works in a series of stages:

1. Concept creation: The development and creation of new ideas through the use of breakthrough thinking's innovative methods of creativity.
2. Concept focus: The development of a target, which includes keeping your organization focused on core values and a core competency. Then, utilizing the creativity generated by concept creation, a set of targets is established using World Class Management, and a road map is developed helping us to achieve the targets.
3. Concept engineering: This is the engineering of the ideas, converting the fuzzy concepts into usable, consumer-oriented ideas. TQM, through the use of a focused, chartered team and through a managed systematic problem solving (SPS) process, helps us to manage the concept from idea to product.
4. Concept in: This is the process of creating a market for the new concept. We transform the concept into a product, service, or system using World Class Management techniques. We may utilize breakthrough thinking to help us develop a meaningful and effective market strategy.
5. Concept management: Both the management of the new concepts and a change in the management approach (management style) is effected by the new concept. Concept management integrates the first four stages of the process (creation, focus, engineering, and in).

For more detail on the concept management process, please read the book *Making Innovation Happen: Concept Management Through Integration.*

2   Breakthrough Thinking

We're getting closer to the end of this chapter and the end of the definitions of the various CPI change management methodologies. At this point we move away from TPS-based tools. Table 3.11 shows what topics are left for us to discuss.

Breakthrough thinking (BT) stresses that in order to solve difficult problems and analyze new opportunities hoping to find creative solutions, our present thinking paradigm must change. Gerald Nadler and Shozo Hibino published *Breakthrough Thinking* in 1990 and *Creative Solution Finding* in 1993 (both published by Prima Lifestyles). In these two books they defined a Japanese-developed paradigm shift in thinking that they called "breakthrough thinking." From a historical viewpoint, our thinking paradigms have been continuously shifting over time. Our conventional thinking paradigm (Descartes thinking) is out of date with a rapidly changing world and needs to shift again to a new thinking paradigm. In the twenty-first century, we have to be multi-thinkers who are able to use three thinking paradigms: God thinking, conventional (Descartes) thinking, and breakthrough thinking.

God thinking focuses on making decisions based on God's will. For some decisions, there is no need for analysis. Behavior is firmly dictated by God's will, our value systems, and our life philosophies. For example, moral

Table 3.11   Quality Methodology

2	Breakthrough Thinking
3	TOC—Theory of Constraints, Bottleneck Analysis, Constraint Analysis
4	Process Reengineering
5	ISO Standards
6	Strategic Mapping
6.1	Hoshin Planning
7	Simulation/Modeling

or ethical issues are decided and are not open for discussion. Conventional thinking starts with an analysis process that focuses on fact or truth finding. When we make a decision, our behavior is based on the facts or on scientific truth. We need the facts in order to make our decisions. Breakthrough thinking starts with the ideal or ultimate objective. When we make a decision, we base our behavior on this objective.

The three thinking paradigms are completely different, and each has a different approach. We have to select and utilize each of these paradigms on case-by-case basis. Thus, someone who uses and intertwines these thinking paradigms is referred to as a "multi-thinker."

Since there is no future that continues along the same lines as our past and present (because of the drastic changes going on in the world), we cannot find futuristic solutions based on past and present facts. Our thinking needs to move away from facts and refocus on the substance, essence, or ideal.

To identify the substance of things is not easy. We have to transform ourselves from having a conventional machine view to a systems-oriented view. The traditional perspective of conventional thinking is to view things as a reductionistic machine, breaking everything down into elemental parts, and neglecting the "whole" organic view.

Breakthrough thinking suggests that "everything is a system," which focuses BT on a holonic view." If we define everything as a system, then everything is a "Chinese box," which means that a bigger box (system) includes a series of smaller boxes (systems). A small box (system) contains still smaller boxes (systems) and so on. Each box (system) has its purpose(s). If you repeatedly ask "What is the purpose?" and then "What is the purpose of that purpose?" and then "What is the purpose of that purpose of that purpose?" and so on, you can reach the biggest box, which is "wholeness." You can view everything from the perspective of this wholeness. BT calls this search the "purpose expansion."

Thus, breakthrough thinking consists of a thinking paradigm and thinking process. The thinking paradigm of breakthrough thinking is the opposite of the paradigm of the conventional thinking. Its main points are expressed as seven principles:

1. Uniqueness principle: Always assume that the problem, opportunity, or issue is different. Don't copy a solution or use a technique from

elsewhere just because the situation may appear to be similar. In using this principle, we have to think about the locus or solution space of the problem. This locus is defined using three points:

    a. Who are the major stakeholders? Whose viewpoint is most important?

    b. What is the location?

    c. When (What is the timing)?

2. Purposes principle: Explore and expand purposes in order to understand what really needs to be accomplished and to identify the substance of things. You can tackle any problem, opportunity, or issue by expanding purposes if you change your epistemology to a systems view. Understanding the context of purposes provides the following strategic advantages:

    a. Pursue the substance of things: We can identify the most essential focus purpose or the greater purpose, often referred to as the substance (core element), of things by expanding purposes.

    b. Work on the right problem or purpose: Focusing on the right purposes helps strip away nonessential aspects to avoid working on just the visible problem or symptom.

    c. Improve the ability to redefine: Redefining is usually difficult. Once you've redefined, you can have different viewpoints, each of which enables you to solve problems from different directions

    d. Eliminate purpose/function(s): From systems theory we learn that a bigger purpose may eliminate a smaller purpose. By focusing on the bigger purpose, you can eliminate unnecessary work/systems/parts, which means that you can get more effective solutions.

    e. More options, more creativity: If you have a purpose hierarchy, you have a lot of alternative solutions.

    f. Holonic view: Take a "big picture" perspective.

3. Solution-after-next (SAN) principle: Design futuristic solutions for the focus purpose and then work backward. Consider the solution you would recommended if in three years you had to start all over. Make changes today based on what might be the solution of the future. Learn from the futuristic ideal solution for the focus purpose, and don't try learning from the past and present situations.

4. Systems principle: Everything we seek to create and restructure is a system. Think of solutions and ideas as systems. When you see everything is a system, you have to consider the eight elements of a system in order to identify the solution:

   a. Purpose—mission, aim, need
   b. Input—people, things, information
   c. Output—people, things, information
   d. Operating steps—process and conversion tasks
   e. Environment—physical and organizational
   f. Human enablers—people, responsibilities, skills to help in the operating steps
   g. Physical enablers—equipment, facilities, materials to use in the operating steps
   h. Information enablers—knowledge, instructions

5. Needed information collection principle: Collect only the information that is necessary to continue the solution finding process. Know your purposes for collecting data and/or information. Study the solutions, not the problems.

6. People design principle: Give everyone who will be affected by the solution or idea the opportunity to participate throughout the process of its development. A solution will work only if people know about it and help to develop and improve it.

7. Betterment timeline principle: Install changes with built-in seeds of future change. Know when to fix it before it breaks. Know when to change it.

The BT process is a reasoning approach toward a situation-specific solution and a design approach. It is an iterative, simultaneous process of mental responses based on the purpose-target-results (PTR) approach. PTR's three phases are

1. Purpose: Identifying the right solution by finding focus purposes, values, and measures

2. Target: Targeting the solution of tomorrow (ideal SAN vision and target solution)

3. Result: Getting and maintaining results toward implementation and systematization

For a more detailed discussion, please read *Breakthrough Thinking*.

3   TOC—Theory of Constraints, Bottleneck Analysis, Constraint Analysis

Theory of constraints (TOC) is a tool used for increasing throughput when a constraint or bottleneck exists in the system. It is based on Goldratt's approach of optimizing bottleneck work centers using work queues around the bottleneck so that there is no idle time within the bottleneck. Because TOC focuses on optimizing a bottleneck, the organization's entire scheduling process is restructured around this bottleneck. The TOC approach was introduced by Dr. Eliyahu M. Goldratt in his 1984 book titled *The Goal* (North River Press). The TOC process seeks to identify the constraint and restructure the rest of the organization around it, through the use of five focusing steps.

The underlying premise of TOC is that organizations should be measured using three factors: throughput, operational expense, and inventory. Throughput is the rate at which the system creates money through sales. Inventory is the dollar value of the resources that the system has tied up and intends to sell. Operational expense is the cost of turning inventory into throughput.

In TOC, only by increasing flow through the bottleneck or constraint can overall throughput be increased. Using the organizational goal of "making money," the five focusing steps are as follows:

1. Identify the constraint or bottleneck. It is the resource or policy that prevents the organization from achieving more of the goal.
2. Determine how to exploit the constraint by maximizing its capacity.
3. Subordinate all other processes in the organization to the constraint so that they support the throughput of the constraint rather than limit it.
4. Elevate the constraint by making whatever changes are necessary to reduce or eliminate the constraint.
5. If, as a result of these steps, a different part of the organization rises to the surface as a new constraint, return to Step 1.

The goal for most supply chains is to optimize the flow of inventory. The TOC tool is effective when used to address supply chain constraints. TOC facilitates a competitive edge based on its ability to reduce the damage caused when the flow of goods is interrupted by shortages and surpluses.

4 Process Reengineering

Business process reengineering is the "rapid and radical" analysis and redesign of workflows and processes within an organization. Business process reengineering is also known as business process redesign, business transformation, or business process change management. Business process reengineering (BPR) began to help organizations rethink how they work in order to dramatically improve customer service, cut operational costs, and become world class competitors.

BPR redesigns the way work is done. It starts with an assessment of the organization's mission, strategic goals, and customer needs. Only after the organization rethinks if it is doing the right things does it go on to decide how best to do it.

Next reengineering focuses on the organization's business processes. It promotes a structured ordering of work steps. The business process can then be decomposed into its specific activities and improved. The process can also be completely redesigned or eliminated altogether.

5 ISO Standards

The International Organization for Standardization (ISO) is an international standards body composed of representatives from various nations. The organization develops and publishes worldwide proprietary industrial and commercial standards. ISO's main products are the International Standards, including published Technical Reports and Technical Specifications. These reports and specifications have become the quality standard for documenting process quality and performance.

6 Strategic Mapping

Strategic mapping is the development of a strategic plan that defines the goals and objectives of an organization. These goals and objectives are then used to define quality performance standards. There are hundreds of formats for strategic maps. One example listed here incorporates the initial high-level elements of an executable strategy map (see Figure 3.25). A map incorporates the following (using the layout from Figure 3.14):

▲ Core competencies
▲ The vision statement

Front-End Strategic Information			
Core Competencies:			
Vision:			
Mission:			
Defining the Customer:			
The Strategy Map			
Priorities and Goals:	**Objectives/Strategies:**	**Metrics:**	**Tasks/Action Items:**
Priority Statement: **End State Statement:**	Objective/Strategy Statement:	Metrics Statements:	Tasks/Action Item Statements:
Goal Statement:	Objective/Strategy Statement:	Metrics Statements:	Tasks/Action Item Statements:
	Objective/Strategy Statement:	Metrics Statements:	Tasks/Action Item Statements:

Figure 3.25   The strategy map.

- ▲ The mission statement
- ▲ The customer
- ▲ The corporate priorities
- ▲ The end state for each priority
- ▲ The goals for each priority

6.1   Hoshin Planning

Hoshin planning, also known as Hoshin Kanri or policy deployment, is a strategy tool devised to capture strategic goals and integrate them with insight about the future so that a plan can be developed that will bring these goals into reality. Hoshin planning is a strategic planning/strategic management methodology introduced by Professor Kaoru Ishikawa in his book *What Is Total Quality Control?* (Prentic Hall Trade) in the late 1950s. He said that each person is the expert in his or her own job. Japanese TQC is designed to use the collective thinking of all employees to make their organization the best. In Professor Ishikawa's words, "Top managers and middle managers must be bold enough to delegate as much authority as possible. That is the way to establish respect for humanity as your management philosophy. It is a management system in which all employees participate, from the top down and from the bottom up, and humanity is fully respected." Adaptations of the concept include the PDCA cycle.

The discipline of Hoshin Kanri helps an organization

▲ Focus on a collaborative, shared goal
▲ Communicate that goal across the organization
▲ Involve everyone in planning to achieve the goal
▲ Hold all participants accountable for achieving their part of the plan

It assumes daily controls and performance measures are in place.

7 Simulation/Modeling

Simulation and modeling are used to imitate some real thing, state of affairs, or process. The act of simulating or modeling something requires representing certain key characteristics or behaviors of a selected system using models or mathematical representations.

Simulation is used to show the eventual real effects of alternative conditions or actions. Simulation is also used when the real system is not available or because it may destroyed by the process. It may also be dangerous or unacceptable to engage in, or it may simply not exist.

Key issues in simulation and modeling include collecting accurate data about the key characteristics and behaviors. The simulation is no better than the accuracy of the approximations and assumptions within it.

Summary

This chapter started us on a journey of never-ending alternatives. There are numerous CPI alternatives, each claiming to be the successful road to quality. We learned that many of these CPI methodologies are interrelated subsets of each other. We also learned that it really does matter which tools we use, because each tool eventually leads us to a different quality destination.

This chapter listed the current CPI fads. It is structured by group, since many of the tools are not really unique tools, but are often subsets or sub-elements of other tools. In this chapter each of the tools was defined and briefly explained.

From here we will press forward by taking this list and analyzing some of the specific characteristics. Then we will map these characteristics to various examples of quality problems. And we will map back the quality problems to a suggested set of tools that can be used to solve each problem.

CHAPTER 4

What Are the Characteristics of Each Option?

Inaction breeds doubt and fear. Action breeds confidence and courage. If you want to conquer fear, do not sit home and think about it. Go out and get busy.

DALE CARNEGIE

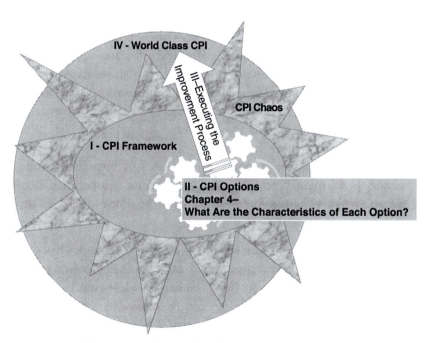

Figure 4.1 The Quality Chaos.

This chapter starts to categorize the tools in ways that will be meaningful later when we try to solve specific problems. Categorization becomes challenging, primarily because there are so many different ways to look at and evaluate quality tools. To start with, let's look at and define the following categories:

- **Quality System** A quality system is a methodology that integrates the quality culture of an organization with the quality tools that it utilizes. A quality system becomes a way of thinking throughout the organization. This is exemplified at Toyota with its TPS or at GE and Motorola with their Six Sigma programs. Everyone, starting with top management and ending at the shop floor, absorbs themselves in this way of thinking and makes it a part of corporate life.
- **Quality Tool** A quality tool is a quality hammer and a nail used to solve a specific problem. It doesn't concern itself with the philosophy of quality.
- **Quality Philosophy** A quality philosophy is a way of thinking. For example, continuously looking for opportunities for quality improvements or providing a quality-based reward mechanism.
- **Acceptance Tool** An acceptance tool is a preparatory tool, helping the organization get ready for a CPI quality change process. It looks at the organization's readiness and willingness for change, both at an individual level and at an organizational level. And it looks at the team dynamics of the organization that will be analyzing and implementing the change.
- **Technical Tool** A technical tool is a technical process or procedure used to implement the change. It involves analyzing the change and then implementing it.
- **Control Tool** A control tool is a quality methodology that monitors quality performance and then offers feedback on that performance. It doesn't make a change. Rather, it informs the user of the need for a change.

Looking at Table 4.1 we can see the quality methodologies listed and graded using each of these definitions. Note that there is often overlap and that a specific methodology might be listed under several categories. This occurs because the methodology can be applied in several different ways. For example, Lean is a system, in that it needs to become part of the corporate way of thinking; it is a bag of tools, as can be seen from the

Table 4.1 Quality Methodology

		Quality System	Quality Tool	Quality Philosophy	Acceptance Tool	Technical Tool	Control Tool
1	TPS—Toyota Production Systems	*		*			
1.1	TQM—Total Quality Management	*		*			
1.1.1	TQM—Deming Version		*	*		*	*
1.1.2	TQM—Crosby Version		*	*		*	*
1.1.3	TQM Juran Version		*	*		*	*
1.2	TQC—Total Quality Control		*			*	*
1.3	Lean	*	*	*			
1.3.1	Process, Project, or Event Charter		*			*	*
1.3.2	Scan		*		*	*	
1.3.3	Kaizen Events, Rapid Improvement Events (RIE)		*	*		*	*
1.3.3.1	PDCA—Plan Do Check Act			*		*	*
1.3.3.2	A3 Reporting		*			*	*
1.3.3.2.01	Root Cause Analysis		*			*	*
1.3.3.3	Acceptance Change Management Tools			*	*		
1.3.3.3.01	Change Acceleration Process (CAP) Model		*	*	*		*

(Continued)

Table 4.1 Quality Methodology (Continued)

		Quality System	Quality Tool	Quality Philosophy	Acceptance Tool	Technical Tool	Control Tool
1.3.3.3.02	Kotter Change Model		*	*	*		*
1.3.3.3.03	Myers Briggs		*		*		
1.3.3.3.04	JoHari Window		*		*		
1.3.3.3.05	Ladder of Inference		*		*		
1.3.3.3.06	PAPT		*		*		
1.3.3.3.07	Situational Leadership		*		*		
1.3.3.4	Technical Tools			*			*
1.3.3.4.01	Value Stream Mapping		*			*	
1.3.3.4.02	Seven Wastes		*	*		*	
1.3.3.4.03	System Flow Chart		*			*	
1.3.3.4.04	Swimlane Chart		*			*	
1.3.3.4.05	Spaghetti Chart		*			*	
1.3.3.4.06	5S		*			*	
1.3.3.4.07	Cellular Work Design		*			*	
1.3.3.4.08	JIT—Just-in-Time		*	*		*	
1.3.3.4.09	Agile		*			*	
1.3.3.4.10	Poka-Yoke (Error Proofing/ Mistake Proofing)		*			*	
1.3.3.4.11	PQ Analysis		*			*	
1.3.3.4.12	TPM—Total Product (Productive, Preventative, or Production) Maintenance		*	*		*	

(Continued)

Table 4.1 Quality Methodology (Continued)

		Quality System	Quality Tool	Quality Philosophy	Acceptance Tool	Technical Tool	Control Tool
1.4.1	DMAIC			*			*
1.4.2	SQC—Statistical Quality Control		*			*	
1.4.3	SPC—Statistical Process Control		*			*	
1.4.4	Queuing Theory		*			*	*
1.4.5	DFSS			*			
1.5	QFD—Quality Functional Deployment	*	*	*		*	*
1.6	Concept Management	*	*	*		*	*
2	Breakthrough Thinking	*	*	*	*	*	*
3	TOC—Theory of Constraints, Bottleneck Analysis, Constraint Analysis	*	*	*		*	*
4	Process Reengineering	*	*	*		*	*
5	ISO Standards	*	*	*		*	*
6	Strategic Mapping	*		*			
6.1	Hoshin Planning	*		*			
7	Simulation/Modeling	*	*	*		*	

long list of support tools that follow after point 1.3 in the table; and it is a philosophy, in that it changes the way we think. Lean users can no longer walk into a McDonald's without immediately looking at the process and mentally trying to fix it. Lean gets into your blood.

Table 4.1 lists all the quality methodologies that were defined in Chapter 3 and categorizes them based on the aforementioned definitions.

One good table always leads to another, and Table 4.2 applies more categorizations to our set of quality methodologies. In this table we look at the following categories:

- ▲ **Contains Specific Metrics**   Some quality methodologies do not care what improvement metrics you utilize, while others are extremely specific about the choice of metrics. If the methodology has a specific metric requirement, then this box will be starred.
- ▲ **Contains Implied Metrics**   This column places a star in the box of any methodology that hints at the importance of and use of metrics, but does not necessarily specify which metrics should be used.
- ▲ **Process/Systems Oriented**   In this box we are looking at quality improvement methodologies that focus on the process and its flow. These methodologies are looking for improvement opportunities within a process.
- ▲ **Targets Variability**   This box marks those tools that look for variability in the process. They are generally statistics-oriented and collect a lot of data from which performance variability can be tracked.
- ▲ **Targets Process Flow**   This box is for methodologies that specifically try to improve the process or flow of work and are not necessarily looking for quality errors.
- ▲ **Top Leadership Involved**   In this box we mark those methodologies that require the participation of top leadership in order to be implemented. Of course, all quality improvements work better with top management commitment, but this box specifically targets those items that need close and personal top management involvement.

A last categorization table will focus on metrics. Table 4.3 explores the long-term and short-term metrics in the following categories:

- ▲ **Customer Satisfaction**   All quality roads lead to customer satisfaction, but the purpose of this box is to identify those quality methodologies that specifically target customer satisfaction as one of their chief goals.

Table 4.2 Quality Methodology

		Contains Specific Metrics	Contains Implied Metrics	Process/ Systems Oriented	Targets Variability	Targets Process Flow	Top Leadership Involved
1	TPS—Toyota Production Systems		*	*	*	*	*
1.1	TQM—Total Quality Management		*				*
1.1.1	TQM—Deming Version		*		*		*
1.1.2	TQM—Crosby Version		*				*
1.1.3	TQM—Juran Version	*			*		*
1.2	TQC—Total Quality Control		*		*		
1.3	Lean		*	*		*	*
1.3.1	Process, Project, or Event Charter	*		*		*	*
1.3.2	Scan			*		*	*
1.3.3	Kaizen Events, Rapid Improvement Events (RIE)			*		*	*
1.3.3.1	PDCA—Plan Do Check Act					*	
1.3.3.2	A3 Reporting	*		*		*	
1.3.3.2.01	Root Cause Analysis			*		*	
1.3.3.3	Acceptance Change Management Tools			*			
1.3.3.3.01	Change Acceleration Process (CAP) Model			*			

Code	Tool					
1.3.3.3.02	Kotter Change Model				*	
1.3.3.3.03	Myers Briggs			*	*	
1.3.3.3.04	JoHari Window			*	*	
1.3.3.3.05	Ladder of Inference			*	*	
1.3.3.3.06	PAPT			*	*	
1.3.3.3.07	Situational Leadership			*	*	
1.3.3.4	Technical Tools	*		*	*	*
1.3.3.4.01	Value Stream Mapping	*		*	*	*
1.3.3.4.02	Seven Wastes	*		*		*
1.3.3.4.03	System Flow Chart	*		*	*	*
1.3.3.4.04	Swimlane Chart	*		*	*	
1.3.3.4.05	Spaghetti Chart	*		*	*	
1.3.3.4.06	5S		*	*		
1.3.3.4.07	Cellular Work Design	*		*	*	
1.3.3.4.08	JIT—Just-in-Time	*		*	*	
1.3.3.4.09	Agile	*		*	*	
1.3.3.4.10	Poka-Yoke (Error Proofing/Mistake Proofing)	*		*	*	*
1.3.3.4.11	PQ Analysis	*	*	*	*	*
1.3.3.4.12	TPM—Total Product (Productive, Preventative, or Production) Maintenance		*	*	*	
1.3.3.4.13	Visual Workplace			*	*	

(Continued)

Table 4.2 Quality Methodology (*Continued*)

		Contains Specific Metrics	Contains Implied Metrics	Process/ Systems Oriented	Targets Variability	Targets Process Flow	Top Leadership Involved
1.3.3.4.14	DFM—Design for Manufacturability	*		*		*	
1.3.3.4.15	SMED—Single Minute Exchange of Die, Quick Changeover	*		*	*	*	
1.3.3.4.16	Kanban	*		*		*	
1.3.3.4.17	Jidoka		*	*	*	*	
1.3.3.4.18	Standard Work		*	*		*	
1.3.3.4.19	Brainstorming			*		*	
1.3.3.4.20	Fishbone Charting			*		*	
1.3.3.4.21	Eight-Step Problem Solving		*	*		*	
1.3.3.4.22	SWOT—Strengths, Weaknesses, Opportunities, Threats			*		*	*
1.3.3.4.23	VOC—Voice of the Customer			*		*	*
1.3.3.4.24	Gemba Walk – Go and See			*		*	
1.3.3.4.25	Gap Analysis		*	*	*	*	
1.3.3.4.26	Five Whys			*		*	
1.3.3.4.27	SIPOC or COPIS			*		*	
1.3.3.4.28	Pull Signaling		*	*		*	

Table 4.3 Quality Methodology/Metrics

		Long Term					Short Term				Not Tied to a Specific Metric
		Customer Satisfaction	Strategic Metrics	Inventory/ Operating Exp	Customer On-time Delivery	Quality Units Produced	Statistical Measures	Cycle Time/ Throughput	Touch Points	Product Lost or Damaged	
1	TPS—Toyota Production Systems										*
1.1	TQM—Total Quality Management	*				*	*				
1.1.1	TQM—Deming Version	*				*	*				
1.1.2	TQM—Crosby Version	*				*					
1.1.3	TQM—Juran Version	*				*	*				
1.2	TQC—Total Quality Control					*	*				
1.3	Lean	*				*	*	*	*		
1.3.1	Process, Project, or Event Charter		*								*
1.3.2	Scan										*
1.3.3	Kaizen Events, Rapid Improvement Events (RIE)										*
1.3.3.1	PDCA—Plan Do Check Act										*
1.3.3.2	A3 Reporting										*
1.3.3.2.01	Root Cause Analysis										*
1.3.3.3	Acceptance Change Management Tools										*

ID	Tool					
1.3.3.3.3.01	Change Acceleration Process (CAP) Model					*
1.3.3.3.3.02	Kotter Change Model					*
1.3.3.3.3.03	Myers Briggs					*
1.3.3.3.3.04	JoHari Window					*
1.3.3.3.3.05	Ladder of Inference					*
1.3.3.3.3.06	PAPT					*
1.3.3.3.3.07	Situational Leadership					*
1.3.3.4	Technical Tools					*
1.3.3.4.01	Value Stream Mapping	*			*	
1.3.3.4.02	Seven Wastes	*			*	*
1.3.3.4.03	System Flow Chart			*	*	*
1.3.3.4.04	Swimlane Chart			*	*	
1.3.3.4.05	Spaghetti Chart	*		*	*	
1.3.3.4.06	5S	*				
1.3.3.4.07	Cellular Work Design	*	*	*	*	
1.3.3.4.08	JIT—Just-in-Time	*	*	*		
1.3.3.4.09	Agile	*	*	*	*	
1.3.3.4.10	Poka-Yoke (Error Proofing/Mistake Proofing)	*	*	*	*	
1.3.3.4.11	PQ Analysis	*	*		*	
1.3.3.4.12	TPM—Total Product (Productive, Preventative, or Production) Maintenance	*	*		*	

(Continued)

Table 4.3 Quality Methodology/Metrics (Continued)

	Long Term					Short Term				Not Tied to a Specific Metric
	Customer Satisfaction	Strategic Metrics	Inventory/ Operating Exp	Customer On-time Delivery	Quality Units Produced	Statistical Measures	Cycle Time/ Throughput	Touch Points	Product Lost or Damaged	
1.3.3.4.13 Visual Workplace						*	*	*		*
1.3.3.4.14 DFM—Design for Manufacturability					*	*	*	*	*	
1.3.3.4.15 SMED—Single Minute Exchange of Die, Quick Changeover						*	*	*		
1.3.3.4.16 Kanban			*		*	*			*	
1.3.3.4.17 Jidoka			*		*	*			*	
1.3.3.4.18 Standard Work					*	*		*	*	
1.3.3.4.19 Brainstorming										*
1.3.3.4.20 Fishbone Charting					*				*	
1.3.3.4.21 Eight-Step Problem Solving										*
1.3.3.4.22 SWOT—Strengths, Weaknesses, Opportunities, Threats										*
1.3.3.4.23 VOC—Voice of the Customer	*			*	*					
1.3.3.4.24 Gemba Walk—Go and See										*
1.3.3.4.25 Gap Analysis										*
1.3.3.4.26 Five Whys										*
1.3.3.4.27 SIPOC or COPIS										*

1.3.3.4.28	Pull Signaling
1.3.3.4.29	Affinity Diagrams
1.3.4	Shingo Prize
1.4	Six Sigma
1.4.1	DMAIC
1.4.2	SQC—Statistical Quality Control
1.4.3	SPC—Statistical Process Control
1.4.4	Queuing Theory
1.4.5	DFSS
1.5	QFD—Quality Functional Deployment
1.6	Concept Management
2	Breakthrough Thinking
3	TOC—Theory of Constraints, Bottleneck Analysis, Constraint Analysis
4	Process Reengineering
5	ISO Standards
6	Strategic Mapping
6.1	Hoshin Planning
7	Simulation/Modeling

▲ **Strategic Metrics** In this column we are looking for methodologies that require metrics that were strategically developed, and not metrics that are operational.

▲ **Inventory/Operation Expense** In this box we are looking for methodologies that specifically target either of the operational measures of inventory improvement or operating expense reduction.

▲ **Customer On-Time Delivery** In this box we are looking for methodologies that focus on the measurement of customer on-time delivery.

▲ **Quality Units Produced** In this box we focus on the throughput metric—specifically, quality throughput. If the methodology specifies this as a required metric, the box will be starred.

▲ **Statistical Measures** This column is for those methodologies that utilize data to generate statistical performance measures.

▲ **Cycle Time/Throughput** In this column we are looking for methodologies that focus on measuring process cycle time or process throughput.

▲ **Product Lost or Damaged** Here the focus is on parts that are damaged or lost, usually in the logistics process. The movement of the parts is causing failures.

▲ **Not a Specific Metric** This is a catch-all category covering all those methodologies that do not focus specifically on metrics.

Integration

Hopefully at this point you're not feeling too confused. The important lesson to learn is that none of these tools should be thought of as an isolated methodology. As you can see in Figure 4.2, there is a tremendous amount of overlap. The lower-level methodologies can be integrated into any number of higher-level tools. For example, the charter (1.3.1) or root cause analysis (1.3.3.2.01), or brainstorming (1.3.3.4.19) are often used in Lean (1.3), Six Sigma (1.4), TQM (1.1), TOC (3), or any number of other high-level methodologies. Nearly all of these tools blend nicely together. At worst, you'll generate some duplication, but that's better than missing something important. Do not feel restricted by the arbitrary numbering system that I created here. Build yourself a combination of tools that will accomplish the objectives you are trying to achieve.

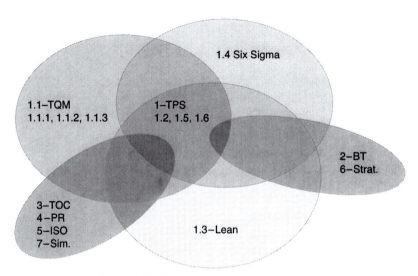

Figure 4.2   Quality Methodology Overlaps.

Summary

This chapter has taken all of the methodologies that were identified and defined in Chapter 3 and has developed some classification categories for them. The purpose of this was to offer differentiating characteristics that will then be used to tie the methodology to the problem. In the next chapter we will see how these classifications are use to make the problem-to-methodology connection.

CHAPTER 5

Mapping CPI Characteristics Against Your Problem

Failure isn't falling down, it's staying down.

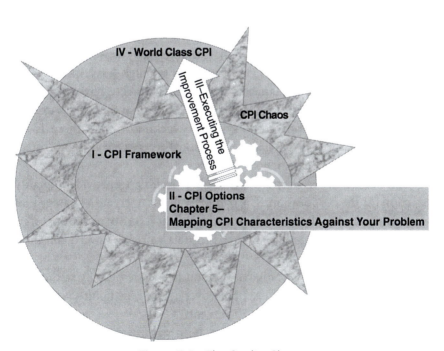

Figure 5.1 The Quality Chaos.

The last couple of chapters have focused on quality methodologies. This chapter starts the discussion of quality problems. To do this, we will look at seven specific examples of quality problems. Each of these problems was drawn from the actual experience of the author. These are real problems with real solutions.

The procedure will be to describe the problem, then show the critical characteristics of that problem, and then map the characteristics against the quality methodology characteristics from the last chapter. The seven cases we will be exploring are as follows:

1. Lack of organizational strategic alignment with no focus on quality
2. Materials tracking issues in logistics
3. Medical lab labeling and reporting errors
4. Information technology programming inefficiencies
5. Poor manufacturing control of production activity
6. Attorney's office workflow inefficiencies
7. High quality defect rate in printing process

Case #1: Lack of Organizational Strategic Alignment with No Focus on Quality

The Organization: A corporation that is fractured by organizational silos, each working independently. Corporate goals are high-level, meaningless slogans generated by top management. No one seems to have ownership in the goals, and most employees don't even know that these goals exist.

The Process/System: The system is lacking in that there is no corporate-wide strategy that has organizational acceptance and gives direction to the various departments.

The Problem: A corporate strategy is lacking to provide a quality direction for the corporation. The company needs to create a strategy that has organization-wide direction and ownership. The organizational silos need to be blended together and need to start working toward common goals.

The Metrics/Goals: Nonexistent and needed, especially in the area of quality. Quality numbers are poor (over 14 percent failure rates) and customer satisfaction was below 70 percent.

The Problem Characteristics: The organization needs to build high-level support for an event focused around building a corporate direction and strategy. The results of the corporate strategy should give clear direction in all aspects of the business, including quality. A structure needs to be established that will chain this strategy down through all the levels of the organization. In this section we take the same set of characteristics and metrics developed in Chapter 4 and map them against the case (Case #1). In Table 5.1 we see how this mapping graphs out.

Mapping the Problem Characteristics to the Quality Methodology Characteristics: In this section we pull in the quality methodologies that map against the characteristics and metrics for this case. Table 5.2 shows which methodologies could be implemented to help with the problem described in Case #1.

What Quality Methodology Did the Author Use?:   In this section the author will reveal what methodology he utilized when he encountered this problem and why. For Case #1 the author used Quality Methodology 6, strategic mapping, as described in Chapter 3. It was very effective in bringing the organization together and focusing them on common goals and directions. The author wanted to additionally implement Quality Methodology 6.1, Hoshin planning; however, the organization in this example had an extremely short-sighted perspective, not wanting to plan beyond two to three years, and Hoshin was too complicated for the customer to pursue at that time. The use of TPS tools (Methodology 1) and TQM (Methodology 1.1) was planned for future implementation, but was not considered as the best tool for the strategic mapping exercise.

What Were the Results?:   The strategic exercise was extremely effective in bringing organizational alignment. The strategic alignment and goals were later used to generate numerous Lean and Six Sigma–based CPI processes. However, a critical rule for all CPI activities was that no improvement was allowed unless it had directly demonstrated that it had strategic alignment. The strategic effort became the guiding light and control mechanism for all future quality improvement initiatives.

What Other Quality Methodologies Could Have Been Used?:   If the strategic effort was for a more typical, long-term-focused organization with a view to the future beyond five or ten years, then Hoshin planning (Methodology 6.1) could be used for organizational alignment and synchronization. TQM (Methodology 1.1) can also be used for organizing and motivating the organization, giving it a spirit of continuous change.

Table 5.1 Lack of Organizational Strategic Alignment with No Focus on Quality

Case #1

Characteristics (From Table 4.1)

Quality System	Quality Tool	Quality Philosophy	Acceptance Tool	Technical Tool	Control Tool
		*			

Characteristics (From Table 4.2)

Contains Specific Metrics	Contains Implied Metrics	Process/Systems-Oriented	Targets Variability	Targets Process Flow	Top Leadership Involved
*		*			*

Metrics (From Table 4.3)

Long Term					Short Term				
Customer Satisfaction	Strategic Metrics	Inventory/ Operating Exp	Customer On-time Delivery	Quality Units Produced	Statistical Measures	Cycle Time/ Throughput	Touch Points	Product Lost or Damaged	Not Tied to a Specific Metric
*									

Table 5.2 Lack of Organizational Strategic Alignment with No Focus on Quality

Case # 1	Characteristics (From Table 4.1)	Quality System	Quality Tool	Quality Philosophy	Acceptance Tool	Technical Tool	Control Tool
1	TPS—Toyota Production Systems	*		*			
1.1	TQM—Total Quality Management	*		*			
1.4.5	DFSS			*			
6	Strategic Mapping	*		*			
6.1	Hoshin Planning	*		*			

	Characteristics (From Table 4.2)	Contains Specific Metrics	Quality Tool	Contains Implied Metrics	Process/Systems-Oriented	Targets Variability	Targets Process Flow	Top Leadership Involved
1	TPS—Toyota Production Systems	*		*	*	*	*	*
1.1	TQM—Total Quality Management			*				*
1.4.5	DFSS	*			*		*	
6	Strategic Mapping	*			*			*
6.1	Hoshin Planning	*			*			*

		Long Term					Short Term				
	Metrics (From Table 4.3)	Customer Satisfaction	Strategic Metrics	Inventory/ Operating Exp	Customer On-time Delivery	Quality Units Produced	Statistical Measures	Cycle Time/ Throughput	Touch Points	Product Lost or Damaged	Not Tied to a Specific Metric
1	TPS—Toyota Production Systems										*
1.1	TQM—Total Quality Management	*				*	*				
1.4.5	DFSS										*
6	Strategic Mapping		*								
6.1	Hoshin Planning		*								

Case #2: Materials Tracking Issues in Logistics

The Organization: The company is a large high-tech company that purchases contracted products from a subcontractor, passes them through a logistics system that includes several shipping organizations and redistribution warehouses before the product finally arrives at the door of the customer. Unfortunately, once the product enters the distribution system, it tends to become lost in the system. There are occasional blips of information where data about the product is sent to a central database, but the movement of the product is inefficient, and for the most part the only meaningful piece of data is the actual arrival of the product at the customer location.

The Process/System: The organization wanted to move to some form of an event management system that would track the movement of the product throughout the supply network. Before they implement the event management process, however, the organization wanted to revamp the business process and make sure the methodologies used were relevant and appropriate.

The Problem: The organization wanted to develop and appropriate a future state for the logistics network. Then they wanted to implement an event management software tool that took advantage of the redesigned future state.

The Metrics/Goals: The key metric for this organization was customer satisfaction. Secondary to that are the metrics of on-time customer delivery, product delivery cycle time, and product lost or damaged in the delivery process. A future goal is to reduce the total amount of dollars of inventory in the logistics process.

The Problem Characteristics: In Table 5.3 we see how the mapping of Case #2 looks.

Mapping the Problem Characteristics to the Quality Methodology Characteristics: Table 5.4 shows which methodologies could be implemented to help with the problem described in Case #2. It's interesting to see that there are no exact matches to the requirements of Case #2, so we will be forced to use a combination of tools in order to meet all the requirements. In this case, I have listed all the possible solution options, but in future cases, the author will narrow the list down to the few most reasonable examples.

What Quality Methodology Did the Author Use?: The author chose the quality methodologies listed in Table 5.5.

This demonstrates that it may take more than one tool to complete this specific project. The listed collection of tools in Table 5.5 does not imply that the author choose the correct set of tools for all logistics problems.

Table 5.3 Materials Tracking Issues in Logistics

Case # 2	Characteristics (From Table 4.1)	Quality System	Quality Tool	Quality Philosophy	Acceptance Tool	Technical Tool	Control Tool
		*	*	*	*	*	*
	Characteristics (From Table 4.2)	Contains Specific Metrics	Contains Implied Metrics	Process/Systems-Oriented	Targets Variability	Targets Process Flow	Top Leadership Involved
		*	*	*	*	*	*

Metrics (From Table 4.3)	Long Term					Short Term				
	Customer Satisfaction	Strategic Metrics	Inventory/ Operating Exp	Customer On-time Delivery	Quality Units Produced	Statistical Measures	Cycle Time/ Throughput	Touch Points	Product Lost or Damaged	Not Tied to a Specific Metric
	*	*	*	*	*	*	*	*	*	*

Table 5.4 Materials Tracking Issues in Logistics

Case # 2	Characteristics (From Table 4.1)	Quality System	Quality Tool	Quality Philosophy	Acceptance Tool	Technical Tool	Control Tool
1.1.3	TQM—Juran version		*	*		*	*
1.3	Lean	*	*	*			*
1.3.1	Process, Project, or Event Charter		*			*	*
1.3.2	Scan		*		*	*	
1.3.3	Kaizen Events, Rapid Improvement Events (RIE)		*			*	*
1.3.3.1	PDCA—Plan Do Check Act			*			*
1.3.3.2	A3 Reporting		*			*	*
1.3.2.01	Root Cause Analysis		*			*	
1.3.3.3	Acceptance Change Management Tools			*	*		
1.3.3.3.01	Change Acceleration Process (CAP) Model		*	*	*		*
1.3.3.3.02	Kotter Change Model		*	*	*		*
1.3.3.3.03	Myers Briggs		*		*		
1.3.3.3.04	JoHari Window		*		*		
1.3.3.3.05	Ladder of Inference		*		*		
1.3.3.3.06	PAPT		*		*		
1.3.3.3.07	Situational Leadership		*		*		
1.3.3.4	Technical Tools			*			
1.3.4.01	Value Stream Mapping		*			*	

Code	Tool			
1.3.3.4.02	Seven Wastes		*	*
1.3.3.4.03	System Flow Chart		*	
1.3.3.4.04	Swimlane Chart		*	
1.3.3.4.05	Spaghetti Chart		*	
1.3.3.4.06	5S		*	
1.3.3.4.07	Cellular Work Design		*	
1.3.3.4.08	JIT—Just-in-Time		*	*
1.3.3.4.09	Agile		*	
1.3.3.4.10	Poka-Yoke (Error Proofing/Mistake Proofing)		*	
1.3.3.4.11	PQ Analysis		*	
1.3.3.4.12	TPM—Total Product (Productive, Preventative, or Production) Maintenance		*	*
1.3.3.4.13	Visual Workplace		*	
1.3.3.4.16	Kanban		*	*
1.3.3.4.17	Jidoka		*	
1.3.3.4.18	Standard Work		*	*
1.3.3.4.19	Brainstorming		*	
1.3.3.4.20	Fishbone Charting		*	
1.3.3.4.21	Eight-Step Problem Solving	*	*	
1.3.3.4.22	SWOT—Strengths, Weaknesses, Opportunities, Threats		*	*

(Continued)

Table 5.4 Materials Tracking Issues in Logistics (Continued)

Case # 2	Characteristics (From Table 4.1)	Quality System	Quality Tool	Quality Philosophy	Acceptance Tool	Technical Tool	Control Tool
1.3.3.4.23	VOC—Voice of the Customer		*			*	
1.3.3.4.24	Gemba Walk—Go and See		*			*	
1.3.3.4.25	Gap Analysis		*			*	
1.3.3.4.26	Five Whys		*			*	
1.3.3.4.27	SIPOC or COPIS		*			*	
1.3.3.4.28	Pull Signaling		*			*	
1.3.3.4.29	Affinity Diagrams		*			*	
1.4	Six Sigma	*	*	*			
1.4.1	DMAIC			*			*
1.4.2	SQC—Statistical Quality Control		*			*	
1.4.3	SPC—Statistical Process Control		*			*	
1.4.4	Queuing Theory		*			*	
1.5	QFD—Quality Functional Deployment	*	*			*	*
1.6	Concept Management	*	*	*	*	*	*
2	Breakthrough Thinking	*	*	*		*	
3	TOC—Theory of Constraints, Bottleneck Analysis, Constraint Analysis	*	*	*		*	*
4	Process Reengineering	*	*	*		*	*
5	ISO Standards	*	*	*		*	*
7	Simulation/Modeling	*	*	*		*	

152

(Continued)

Characteristics (From Table 4.2)		Contains Specific Metrics	Contains Implied Metrics	Process/Systems-Oriented	Targets Variability	Targets Process Flow	Top Leadership Involved
1.1.3	TQM—Juran Version	*			*		*
1.3	Lean		*	*		*	*
1.3.1	Process, Project, or Event Charter	*		*		*	*
1.3.2	Scan			*		*	*
1.3.3	Kaizen Events, Rapid Improvement Events (RIE)			*		*	*
1.3.3.1	PDCA—Plan Do Check Act					*	
1.3.3.2	A3 Reporting	*		*		*	
1.3.3.2.01	Root Cause Analysis			*		*	
1.3.3.3	Acceptance Change Management Tools			*			
1.3.3.3.01	Change Acceleration Process (CAP) Model			*			
1.3.3.3.02	Kotter Change Model			*			
1.3.3.3.03	Myers Briggs			*			
1.3.3.3.04	JoHari Window			*			
1.3.3.3.05	Ladder of Inference			*			
1.3.3.3.06	PAPT			*			
1.3.3.3.07	Situational Leadership			*			
1.3.3.4	Technical Tools	*	*	*	*	*	
1.3.3.4.01	Value Stream Mapping	*		*		*	
1.3.3.4.02	Seven Wastes			*		*	

Table 5.4 Materials Tracking Issues in Logistics (Continued)

	Characteristics (From Table 4.2)	Contains Specific Metrics	Contains Implied Metrics	Process/Systems-Oriented	Targets Variability	Targets Process Flow	Top Leadership Involved
1.3.3.4.03	System Flow Chart	*		*		*	
1.3.3.4.04	Swimlane Chart	*		*		*	
1.3.3.4.05	Spaghetti Chart	*				*	
1.3.3.4.07	Cellular Work Design	*		*		*	
1.3.3.4.08	JIT—Just-in-Time	*		*		*	
1.3.3.4.09	Agile	*		*		*	
1.3.3.4.10	Poka-Yoke (Error Proofing/Mistake Proofing)	*		*	*	*	
1.3.3.4.11	PQ Analysis	*		*	*	*	
1.3.3.4.12	TPM—Total Product (Productive, Preventative, or Production) Maintenance		*	*	*	*	
1.3.3.4.13	Visual Workplace			*		*	
1.3.3.4.14	DFM—Design for Manufacturability		*	*		*	
1.3.3.4.15	SMED—Single Minute Exchange of Die, Quick Changeover	*		*	*	*	
1.3.3.4.16	Kanban	*		*		*	
1.3.3.4.17	Jidoka		*	*	*	*	
1.3.3.4.18	Standard Work		*	*		*	

Item						
1.3.3.4.19	Brainstorming	*		*		*
1.3.3.4.20	Fishbone Charting	*		*		*
1.3.3.4.21	Eight-Step Problem Solving	*	*	*		*
1.3.3.4.22	SWOT—Strengths, Weaknesses, Opportunities, Threats		*	*		*
1.3.3.4.23	VOC—Voice of the Customer	*		*		*
1.3.3.4.24	Gemba Walk—Go and See	*		*		*
1.3.3.4.25	Gap Analysis	*	*	*		*
1.3.3.4.26	Five Whys	*		*		*
1.3.3.4.27	SIPOC or COPIS	*		*		*
1.3.3.4.28	Pull Signaling		*	*		*
1.3.3.4.29	Affinity Diagrams	*		*		*
1.3.4	Shingo Prize	*	*	*		*
1.4	Six Sigma	*	*	*		*
1.4.1	DMAIC			*		*
1.4.2	SQC—Statistical Quality Control	*		*		*
1.4.3	SPC—Statistical Process Control	*		*		*
1.4.4	Queuing Theory	*	*	*		*
1.5	QFD—Quality Functional Deployment		*	*		*
1.6	Concept Management	*	*	*		*
2	Breakthrough Thinking		*	*		

(Continued)

Table 5.4 Materials Tracking Issues in Logistics (Continued)

Characteristics (From Table 4.2)	Contains Specific Metrics	Contains Implied Metrics	Process/Systems-Oriented	Targets Variability	Targets Process Flow	Top Leadership Involved	Not Tied to a Specific Metric
3 TOC—Theory of Constraints, Bottleneck Analysis, Constraint Analysis	*		*		*		
4 Process Reengineering			*		*	*	
5 ISO Standards	*	*	*		*		
7 Simulation/Modeling		*	*	*	*		

Metrics (From Table 4.3)	Long Term					Short Term				Not Tied to a Specific Metric
	Customer Satisfaction	Strategic Metrics	Inventory/ Operating Exp	Customer On-time Delivery	Quality Units Produced	Statistical Measures	Cycle Time/ Through put	Touch Points	Product Lost or Damaged	
1.1.3 TQM—Juran Version	*				*	*				
1.3 Lean	*				*	*	*			
1.3.1 Process, Project, or Event Charter		*								*
1.3.2 Scan										*
1.3.3 Kaizen Events, Rapid Improvement Events (RIE)										*
1.3.3.1 PDCA—Plan Do Check Act										*
1.3.3.2 A3 Reporting										*
1.3.2.01 Root Cause Analysis										*

Index	Tool							
1.3.3.3	Acceptance Change Management Tools	*						
1.3.3.3.01	Change Acceleration Process (CAP) Model	*						
1.3.3.3.02	Kotter Change Model	*						
1.3.3.3.03	Myers Briggs	*						
1.3.3.3.04	JoHari Window	*						
1.3.3.3.05	Ladder of Inference	*						
1.3.3.3.06	PAPT	*						
1.3.3.3.07	Situational Leadership	*						
1.3.3.4	Technical Tools	*						
1.3.3.4.01	Value Stream Mapping				*			
1.3.3.4.02	Seven Wastes		*	*	*		*	*
1.3.3.4.03	System Flow Chart			*	*			*
1.3.3.4.04	Swimlane Chart			*	*			*
1.3.3.4.05	Spaghetti Chart			*	*			
1.3.3.4.06	5S						*	*
1.3.3.4.07	Cellular Work Design		*		*	*	*	*
1.3.3.4.08	JIT—Just-in-Time		*		*	*	*	*
1.3.3.4.09	Agile		*		*	*	*	
1.3.3.4.10	Poka-Yoke (Error Proofing/Mistake Proofing)		*			*	*	*
1.3.3.4.11	PQ Analysis		*			*		*

(Continued)

Table 5.4 Materials Tracking Issues in Logistics (Continued)

		Long Term					Short Term				Not Tied to a Specific Metric
	Metrics (From Table 4.3)	Customer Satisfaction	Strategic Metrics	Inventory/ Operating Exp	Customer On-time Delivery	Quality Units Produced	Statistical Measures	Cycle Time/ Through put	Touch Points	Product Lost or Damaged	
1.3.3.4.12	TPM—Total Product (Productive, Preventative, or Production) Maintenance	*				*				*	
1.3.3.4.13	Visual Workplace				*			*			
1.3.3.4.14	DFM—Design for Manufacturability					*	*	*	*	*	
1.3.3.4.15	SMED—Single Minute Exchange of Die, Quick Changeover							*	*		
1.3.3.4.16	Kanban			*		*	*	*		*	
1.3.3.4.17	Jidoka			*		*		*		*	
1.3.3.4.18	Standard Work					*	*	*	*	*	
1.3.3.4.19	Brainstorming										*
1.3.3.4.20	Fishbone Charting					*				*	
1.3.3.4.21	Eight-Step Problem Solving										*
1.3.3.4.22	SWOT—Strengths, Weaknesses, Opportunities, Threats										*
1.3.3.4.23	VOC—Voice of the Customer	*			*	*					
1.3.3.4.24	Gemba Walk—Go and See										*

1.3.3.4.25	Gap Analysis
1.3.3.4.26	Five Whys
1.3.3.4.27	SIPOC or COPIS
1.3.3.4.28	Pull Signaling
1.3.3.4.29	Affinity Diagrams
1.3.4	Shingo Prize
1.4	Six Sigma
1.4.1	DMAIC
1.4.2	SQC—Statistical Quality Control
1.4.3	SPC—Statistical Process Control
1.4.4	Queuing Theory
1.5	QFD—Quality Functional Deployment
1.6	Concept Management
2	Breakthrough Thinking
3	TOC—Theory of Constraints, Bottleneck Analysis, Constraint Analysis
4	Process Reengineering
5	ISO Standards
7	Simulation/Modeling

Table 5.5 Materials Tracking Issues in Logistics

Case # 2	Tools Used by the Author
1.3	Lean
1.3.1	Process, Project, or Event Charter
1.3.2	Scan
1.3.3	Kaizen Events, Rapid Improvement Events (RIE)
1.3.3.1	PDCA—Plan Do Check Act
1.3.3.2	A3 Reporting
1.3.3.2.01	Root Cause Analysis
1.3.3.3	Acceptance Change Management Tools
1.3.3.3.03	Myers Briggs
1.3.3.4	Technical Tools
1.3.3.4.01	Value Stream Mapping
1.3.3.4.02	Seven Wastes
1.3.3.4.05	Spaghetti Chart
1.3.3.4.18	Standard Work
1.3.3.4.19	Brainstorming
1.3.3.4.20	Fishbone Charting
1.3.3.4.21	Eight-Step Problem Solving
1.3.3.4.23	VOC—Voice of the Customer
1.3.3.4.24	Gemba Walk—Go and See
1.3.3.4.25	Gap Analysis
1.3.3.4.26	Five Whys
1.3.3.4.27	SIPOC or COPIS

However, this collection turned out to be extremely successful for this specific project.

What Were the Results?: The Lean event recognized several needs, which were listed in the gap analysis. These include:

1. An event management and tracking system, which required an IT implementation
2. Inventory management at the various logistics nodes
3. Cycle time compression Lean events were scheduled

The improvement results included a 50+ percent reduction in inventory, a 60+ percent reduction in cycle time, a 75 percent reduction in damaged or delayed shipments, a move in customer satisfaction from the 60s percentage-wise to the low 90s, and a similar increase in on-time product deliveries.

Table 5.6   Materials Tracking Issues in Logistics

Case # 2	Tools Used by the Author		Possible Substitutes
1.3	Lean	1.1.3	TQM—Juran Version
		1.4	Six Sigma
		1.5	QFD—Quality Functional Deployment
		1.6	Concept Management
		4	Process Reengineering
		5	ISO Standards
1.3.1	Process, Project, or Event Charter		
1.3.2	Scan		(could be skipped)
1.3.3	Kaizen Events, Rapid Improvement Events (RIE)		
1.3.3.1	PDCA—Plan Do Check Act	1.4.1	DMAIC
1.3.3.2	A3 Reporting		
1.3.3.2.01	Root Cause Analysis		
1.3.3.3	Acceptance Change Management Tools	1.3.3.4.22	SWOT—Strengths, Weaknesses, Opportunities, Threats
1.3.3.3.03	Myers Briggs	1.3.3.3.04	JoHari Window
		1.3.3.3.05	Ladder of Inference
		1.3.3.3.06	PAPT
		1.3.3.3.07	Situational Leadership
1.3.3.4	Technical Tools		
1.3.3.4.01	Value Stream Mapping	1.4.2	SQC—Statistical Quality Control
		1.4.3	SPC—Statistical Process Control
		1.4.4	Queuing Theory
		3	TOC—Theory of Constraints, Bottleneck Analysis, Constraint Analysis
1.3.3.4.02	Seven Wastes		
1.3.3.4.05	Spaghetti Chart		
1.3.3.4.18	Standard Work		
1.3.3.4.19	Brainstorming	7	Simulation/Modeling
1.3.3.4.20	Fishbone Charting	1.3.3.4.29	Affinity Diagrams
		2	Breakthrough Thinking
1.3.3.4.21	Eight-Step Problem Solving		
1.3.3.4.23	VOC—Voice of the Customer		
1.3.3.4.24	Gemba Walk—Go and See		
1.3.3.4.25	Gap Analysis		
1.3.3.4.26	Five Whys	2	Breakthrough Thinking
1.3.3.4.27	SIPOC or COPIS		

What Other Quality Methodologies Could Have Been Used?: As we can see from Table 5.4, any number of tools could have been substituted to complete this project. For example, in Table 5.6 we see a list of possible substitutions.

Case #3: Medical Lab Labeling and Reporting Errors

The Organization: This organization is a hospital that had numerous problems, but the focus of this exercise was to identify and eliminate lab reporting errors. For example, the labeling of lab samples often contained errors and caused confusion, causing the patient to be recalled for retesting. Samples may be mislabeled or contaminated. The test results were often recorded to the wrong individual. And the testing process was so slow and cumbersome that patients could be left waiting in the emergency room (ER) for unnecessarily excessively long periods, both frustrating the patients and tying up valuable ER bed resources that could be utilized to service other patients.

The Process/System: There wasn't really a system. When a lab sample was taken, it was labeled with a handwritten label and sent off to the lab. Poor handwriting, transcription errors, or smudges could cause the sample to be processed incorrectly. Then, when the lab report was sent back, it could also be confused with incorrectly transcribed information.

The Problem: The biggest problem was a lack of customer satisfaction and ever-increasing medical employee frustration. Even worse could be the incorrect diagnosis based on incorrect lab information. And even the hospital front office got involved because they were frustrated by the poor bed utilization that was caused by lengthy wait times in the ER.

The Metrics/Goals: The goal was a 50 percent reduction in errors and a 25 percent reduction in processing cycle time.

The Problem Characteristics: In Table 5.7 we see how the mapping of Case #3 looks.

Mapping the Problem Characteristics to the Quality Methodology Characteristics: Table 5.8 shows which methodologies could be implemented to help with the problem described in Case #3. Obviously, you wouldn't use all these tools, and it becomes the role of the facilitator to determine which tools to use based on which problems are identified. For example, if the problem seems more process-oriented (which in this case it was), the Lean methodology would be employed. But if the problem

Table 5.7 Medical Lab Labeling and Reporting Errors

Case # 3

Characteristics (From Table 4.1)

Quality System	Quality Tool	Quality Philosophy	Acceptance Tool	Technical Tool	Control Tool
	*		*	*	*

Characteristics (From Table 4.2)

Contains Specific Metrics	Contains Implied Metrics	Process/Systems-Oriented	Targets Variability	Targets Process Flow	Top Leadership Involved
	*	*	*	*	

Metrics (From Table 4.3)

Long Term					Short Term				
Customer Satisfaction	Strategic Metrics	Inventory/ Operating Exp	Customer On-time Delivery	Quality Units Produced	Statistical Measures	Cycle Time/ Throughput	Touch Points	Product Lost or Damaged	Not Tied to a Specific Metric
*		*	*	*	*	*	*	*	

163

Table 5.8 Medical Lab Labeling and Reporting Errors

Case # 3	Analyzed Against Tables 4.1, 4.2, and 4.3
1.1.1	TQM—Deming Version
1.1.2	TQM—Crosby Version
1.1.3	TQM—Juran Version
1.2	TQC—Total Quality Control
1.3	Lean
1.3.1	Process, Project, or Event Charter
1.3.2	Scan
1.3.3	Kaizen Events, Rapid Improvement Events (RIE)
1.3.3.1	PDCA—Plan Do Check Act
1.3.3.2	A3 Reporting
1.3.3.2.01	Root Cause Analysis
1.3.3.3	Acceptance Change Management Tools
1.3.3.3.01	Change Acceleration Process (CAP) Model
1.3.3.3.02	Kotter Change Model
1.3.3.3.03	Myers Briggs
1.3.3.3.04	JoHari Window
1.3.3.3.05	Ladder of Inference
1.3.3.3.06	PAPT
1.3.3.3.07	Situational Leadership
1.3.3.4	Technical Tools
1.3.3.4.01	Value Stream Mapping
1.3.3.4.02	Seven Wastes
1.3.3.4.05	Spaghetti Chart
1.3.3.4.10	Poka-Yoke (Error Proofing/Mistake Proofing)
1.3.3.4.13	Visual Workplace
1.3.3.4.18	Standard Work
1.3.3.4.19	Brainstorming
1.3.3.4.20	Fishbone Charting
1.3.3.4.21	Eight-Step Problem Solving
1.3.3.4.24	Gemba Walk—Go and See
1.3.3.4.25	Gap Analysis
1.3.3.4.26	Five Whys
1.3.3.4.27	SIPOC or COPIS
1.3.3.4.29	Affinity Diagrams
1.4	Six Sigma
1.4.1	DMAIC
1.4.2	SQC—Statistical Quality Control

Case # 3	Analyzed Against Tables 4.1, 4.2, and 4.3
1.4.3	SPC—Statistical Process Control
1.4.4	Queuing Theory
1.4.5	DFSS
3	TOC—Theory of Constraints, Bottleneck Analysis, Constraint Analysis
4	Process Reengineering
7	Simulation/Modeling

seemed to focus on process and measurement variance, then one of the statistical tools should be used, like Six Sigma or statistical process control. Note that I have stopped using the detail characteristics lists because it makes the tables extremely lengthy and it's easy enough to go back and compare them against the charts in Chapter 4. Again, there are no exact matches, so we take a combination of tools in order to meet all the requirements of the solution. The author will only list those that he may reasonably use in this situation.

What Quality Methodology Did the Author Use?:   The author chose the quality methodologies listed in Table 5.9.

Table 5.9   Medical Lab Labeling and Reporting Errors

Case # 3	Tools Used by the Author
1.3	Lean
1.3.1	Process, Project, or Event Charter
1.3.3	Kaizen Events, Rapid Improvement Events (RIE)
1.3.3.2.01	Root Cause Analysis
1.3.3.4	Technical Tools
1.3.3.4.01	Value Stream Mapping
1.3.3.4.02	Seven Wastes
1.3.3.4.05	Spaghetti Chart
1.3.3.4.10	Poka-Yoke (Error Proofing/Mistake Proofing)
1.3.3.4.13	Visual Workplace
1.3.3.4.18	Standard Work
1.3.3.4.20	Fishbone Charting
1.3.3.4.24	Gemba Walk—Go and See
1.3.3.4.25	Gap Analysis
1.3.3.4.26	Five Whys

What Were the Results?: The Lean event recognized several needs, which were listed in the gap analysis. These include:

1. A barcode labeling process, which would minimize the use of hand-written documents, requires an IT implementation
2. A procedural change focused on standard work and mistake proofing
3. Cycle time compression of the various processes

The improvement results included a 70+ percent reduction in errors and a 50+ percent reduction in cycle time.

What Other Quality Methodologies Could Have Been Used?: As we can see from Table 5.8, any number of tools could have been substituted to complete this project. For example, Six Sigma could have been used to evaluate labeling errors and variance discrepancies. DFSS or process reengineering could have been used to develop an entirely new lab sample processing procedure. Or any of the TQM tools could have been used, and many of them would have given us the same performance improvement results.

Case #4: Information Technology Programming Inefficiencies

The Organization: The company is a large multinational consulting company with 108,000 employees, 90,000+ of whom work in IT technology. The concern is that IT has become such a niche, exclusive, status-driven profession that a hint of standardization is often treated as heresy. The result is that there is a lot of overlap and redundancy in the development process. In addition, the process lacks any control on errors.

The Process/System: The process that this effort is focused on is the software development and test process.

The Problem: The problem is that the various IT subunits seem disconnected, each heading off in their own version of best practices, and each defining quality differently. There is no standardization in the process. The same IT development process run in different organizations can have varying levels of quality and have different cycle times.

The Metrics/Goals: The primary goal is to standardize the work process. Secondary goals include eliminating work redundancy, improving work quality, and reducing cycle times.

The Problem Characteristics: In Table 5.10 we see how the mapping of Case #4 looks.

Table 5.10 Information Technology Programming Inefficiencies

Case # 4

Characteristics (From Table 4.1)	Quality System	Quality Tool	Quality Philosophy	Acceptance Tool	Technical Tool	Control Tool
Characteristics (From Table 4.2)	Contains Specific Metrics	Contains Implied Metrics	Process/Systems-Oriented	Targets Variability	Targets Process Flow	Top Leadership Involved
	*	*	*	*	*	*

Metrics (From Table 4.3)	Long Term					Short Term				
	Customer Satisfaction	Strategic Metrics	Inventory/ Operating Exp	Customer On-time Delivery	Quality Units Produced	Statistical Measures	Cycle Time/ Throughput	Touch Points	Product Lost or Damaged	Not Tied to a Specific Metric
	*			*	*	*	*		*	

Mapping the Problem Characteristics to the Quality Methodology Characteristics: Table 5.11 shows which methodologies could be implemented to help with the problem described in Case #4. As always, you wouldn't use all these tools. Again, there are no exact matches. A combination of tools is used in order to meet all the requirements of the solution. The author will only list those that he may reasonably use in this situation.

What Quality Methodology Did the Author Use?: The author chose the quality methodologies listed in Table 5.12.

What Were the Results?: In this example, since it is somewhat unique, the author will offer some additional depth. The IT development company recognized the need for process performance efficiencies in IT and implemented Lean throughout their internal IT process. Later, they

Table 5.11 Information Technology Programming Inefficiencies

Case # 4	Analyzed Against Tables 4.1, 4.2, and 4.3
1.3	Lean
1.3.3.1	PDCA—Plan Do Check Act
1.3.3.2	A3 Reporting
1.3.3.2.01	Root Cause Analysis
1.3.3.4	Technical Tools
1.3.3.4.01	Value Stream Mapping
1.3.3.4.03	System Flow Chart
1.3.3.4.04	Swimlane Chart
1.3.3.4.09	Agile
1.3.3.4.10	Poka-Yoke (Error Proofing/Mistake Proofing)
1.3.3.4.18	Standard Work
1.3.3.4.19	Brainstorming
1.3.3.4.20	Fishbone Charting
1.3.3.4.25	Gap Analysis
1.4	Six Sigma
1.4.1	DMAIC
1.4.5	DFSS
4	Process Reengineering

Table 5.12　Information Technology Programming Inefficiencies

Case # 4	Tools Used by the Author
1.3	Lean
1.3.3.1	PDCA—Plan Do Check Act
1.3.3.2	A3 Reporting
1.3.3.2.01	Root Cause Analysis
1.3.3.4	Technical Tools
1.3.3.4.01	Value Stream Mapping
1.3.3.4.03	System Flow Chart
1.3.3.4.04	Swimlane Chart
1.3.3.4.18	Standard Work
1.3.3.4.19	Brainstorming
1.3.3.4.20	Fishbone Charting
1.3.3.4.25	Gap Analysis

expanded their Lean IT capabilities to include IT external clients.[1] This Lean process has been so successful that it is now an integrated part of their culture. The number of Lean improvement initiatives has skyrocketed, as can be seen in Figure 5.2.

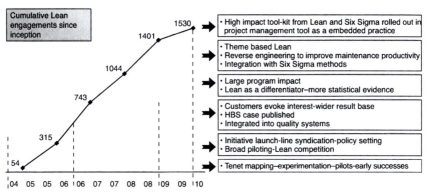

1300 + Lean engagement have been completed so far and 300+ are in progress this year.
Projects that have applied Lean principles have been able to demonstrate significantly differentiated benefits.
10–15% upside on productivity
20% higher capability to handle requirements volatility has been demonstrated across projects
15–20% reduction in delivered defect rate
30% enhanced ability to pull back projects that had adverse performance variances
Projects that have leveraged Lean principles have demonstrated higher consistency of performance
When applied to very large projects, Lean has helped to offset the challenges of complexity and size

Figure 5.2　Lean in IT timeline.

---

[1]　The figures used in this case study come from a Wipro presentation on Lean implementations created by Seema Walunjkar in the Wipro Global Delivery Organization.

The figure denotes the typical challenges in the phases of a software development/maintenance project. It is not specific to any project lifecycle

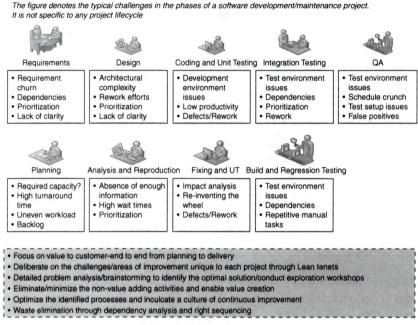

Requirements	Design	Coding and Unit Testing	Integration Testing	QA
• Requirement churn • Dependencies • Prioritization • Lack of clarity	• Architectural complexity • Rework efforts • Prioritization • Lack of clarity	• Development environment issues • Low productivity • Defects/Rework	• Test environment issues • Dependencies • Prioritization • Rework	• Test environment issues • Schedule crunch • Test setup issues • False positives

Planning	Analysis and Reproduction	Fixing and UT	Build and Regression Testing
• Required capacity? • High turnaround time • Uneven workload • Backlog	• Absence of enough information • High wait times • Prioritization	• Impact analysis • Re-inventing the wheel • Defects/Rework	• Test environment issues • Dependencies • Repetitive manual tasks

- Focus on value to customer-end to end from planning to delivery
- Deliberate on the challenges/areas of improvement unique to each project through Lean tenets
- Detailed problem analysis/brainstorming to identify the optimal solution/conduct exploration workshops
- Eliminate/minimize the non-value adding activities and enable value creation
- Optimize the identified processes and inculcate a culture of continuous improvement
- Waste elimination through dependency analysis and right sequencing

Figure 5.3 Applying Lean.

The company was extremely successful with its Lean IT practice. The author is presenting the methodology in the company's own words. Figure 5.3 shows the phases and challenges in the software development life cycle.

Continuing the story, we can see in Figure 5.4 some of the areas of waste that occurred in the IT life cycle.[2]

In addition, Figure 5.5 suggests areas where Lean IT has been exceptionally successful. The diamond areas of the chart show key areas of success.

Figure 5.6 shows the details on how a change management approach for launching Lean throughout the organization works.

Figure 5.7 graphs out the value stream mapping (VSM) process used by this organization.

Here are three examples where Lean IT has been successfully applied with impressive results. In the first, the Lean effort focused on bank account

---

[2] The figures used in this case study come from a Wipro presentation on Lean implementations created by Seema Walunjkar in the Wipro Global Delivery Organization.

Waste Category	Examples (Software)	Waste Category	Examples (Software)
Transport	• Searching for required information (document, email etc.) • Changing requirements, evolving requirements • FTP/Copy	Over-production	• Duplicate test cases • Extra features • Unused features
Inventory	• Frequent task switching results in half-baked inventory and loss of context • Backlog. Over skill	Over-processing	• Redundant reviews, Irrelevant training. Duplicate builds • Obsolete test cases • Duplicate test cases • Unnecessary meetings • For every code drop, every engineer initiates ftp and does a build
Motion	• Customer deliverable going through multiple hands–customer, onsite co-ordinator, offshore team • Frequent travel between locations for reviews • Test setup	Defects and Rework	• Defects • Rework • Poor documentation • Incomplete documentation • Efforts spent in tracing the test setup (Other members disturb the setups to fill the equipment shortage in their setups)
Waiting	• Waiting for customer feedback, information, resources • Waiting for completion of predecessor tasks, clarification on requirements • Delayed reviews		

Figure 5.4   Waste categorization in IT.

Business Challenge	Impact
Application Development	
Consistent quality	◆
Productivity improvement	■
Adherence to Plan	◆
Handling requirement volatility	●
App Maintenance	
Improvement in cycle time	◆
Improvement in productivity	■
Application reengineering	◆
Backlog reduction	■
Production Support	
Improvement in SLA adherence	◆
Backlog reduction	■
Ticket in-flow reduction	■
Optimal capacity planning based on the demand	◆

◆ High
■ Medium
● Low

Figure 5.5   Areas where Lean has worked.

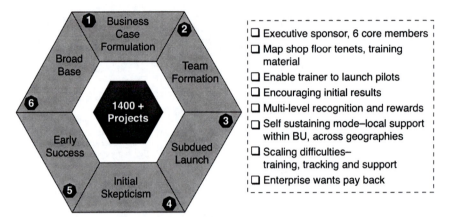

Figure 5.6 Methodology for launching Lean.

interactions with the IT system. The current environment was insufficient in meeting deadlines. The Lean process started with a current state map of user-system interactions and of data/information flow. Using spaghetti charting and system flow charting the team successfully improved overall systems responsiveness, resulting in a 10 percent productivity improvement and extensive error reduction.

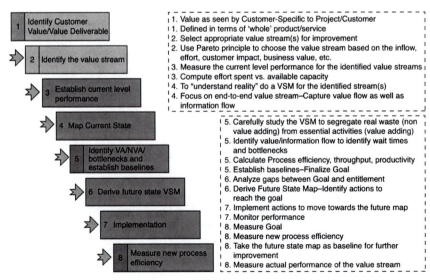

Figure 5.7 Value Stream Mapping.

A second example focused on IT maintenance support in the application of patches for a leading insurance provider experiencing poor quality and a large quantity of post-delivery defects. They now had a support staff correcting 190 to 200 defect incidents monthly. The Lean tools used included VSM and visual controls. The program resulted in a process efficiency improvement going from 11 percent to 29 percent, a 263 percent improvement.

A third example looks at a global healthcare company focused on a large enterprise resources planning (ERP) integration with a focus on discrepancies in the ticket handling process. The CPI process used VSM and incorporated visual controls and mistake-proofing checklists. The improvement resulted in a 69 percent reduction in resolution cycle time and an average 300 percent increase in productivity.

What Other Quality Methodologies Could Have Been Used?:   As we can see from Table 5.11, any number of tools could have been substituted to complete this project. For example, Six Sigma could have been used to evaluate errors and variance discrepancies.

Case #5: Poor Manufacturing Control of Production Activity

The Organization: This is a manufacturing company with 50+ manufacturing facilities worldwide. They produce high-tech products that are resold by OEM (Original Equipment Manufacturing) companies.

The Process/System: They previously had a Manufacturing Execution System (MES) that they were using to drive and control production activity, but they found it to be inaccurate. It caused system delays and process failures because of the poor data-tracking process.

The Problem: They wanted to install a new MES system that would give them real-time information about what was occurring in their production process. They also wanted new functionality because the industry was moving from a build-to-stock toward a build-to-order environment and that would require tight control of the production process.

The Metrics/Goals: The goal was to reduce processing cycle time and to reduce the number of IT touch points. They wanted a system that was as easy to run as possible and that would minimize the number of data errors.

The Problem Characteristics: In Table 5.13 we see how the mapping of Case #5 looks.

Table 5.13 Poor Manufacturing Control of Production Activity

Case # 5	Characteristics (From Table 4.1)					
	Quality System	Quality Tool	Quality Philosophy	Acceptance Tool	Technical Tool	Control Tool
	*				*	

	Characteristics (From Table 4.2)					
	Contains Specific Metrics	Contains Implied Metrics	Process/Systems-Oriented	Targets Variability	Targets Process Flow	Top Leadership Involved
	*		*			

	Metrics (From Table 4.3)									
	Long Term					Short Term				
	Customer Satisfaction	Strategic Metrics	Inventory/ Operating Exp	Customer On-time Delivery	Quality Units Produced	Statistical Measures	Cycle Time/ Throughput	Touch Points	Product Lost or Damaged	Not Tied to a Specific Metric
				*	*		*		*	

Mapping the Problem Characteristics to the Quality Methodology Characteristics: Table 5.14 shows which methodologies could be implemented to help with the problem described in Case #5. The author will only list those that he may reasonably use in this situation.

Table 5.14   Poor Manufacturing Control of Production Activity

Case # 5	Analyzed Against Tables 4.1, 4.2, and 4.3
1.1	TQM—Total Quality Management
1.1.1	TQM—Deming Version
1.1.2	TQM—Crosby Version
1.1.3	TQM—Juran Version
1.3	Lean
1.3.1	Process, Project, or Event Charter
1.3.2	Scan
1.3.3	Kaizen Events, Rapid Improvement Events (RIE)
1.3.3.1	PDCA—Plan Do Check Act
1.3.3.2	A3 Reporting
1.3.3.2.01	Root Cause Analysis
1.3.3.3	Acceptance Change Management Tools
1.3.3.3.01	Change Acceleration Process (CAP) Model
1.3.3.3.02	Kotter Change Model
1.3.3.3.03	Myers Briggs
1.3.3.3.04	JoHari Window
1.3.3.3.05	Ladder of Inference
1.3.3.3.06	PAPT
1.3.3.3.07	Situational Leadership
1.3.3.4	Technical Tools
1.3.3.4.01	Value Stream Mapping
1.3.3.4.02	Seven Wastes
1.3.3.4.03	System Flow Chart
1.3.3.4.04	Swimlane Chart
1.3.3.4.07	Cellular Work Design
1.3.3.4.08	JIT—Just-in-Time
1.3.3.4.09	Agile
1.3.3.4.10	Poka-Yoke (Error Proofing/Mistake Proofing)
1.3.3.4.13	Visual Workplace
1.3.3.4.14	DFM—Design for Manufacturability
1.3.3.4.15	SMED—Single Minute Exchange of Die, Quick Changeover

(Continued)

Table 5.14 Poor Manufacturing Control of Production Activity (*Continued*)

Case # 5	Analyzed Against Tables 4.1, 4.2, and 4.3
1.3.3.4.16	Kanban
1.3.3.4.17	Jidoka
1.3.3.4.18	Standard Work
1.3.3.4.19	Brainstorming
1.3.3.4.20	Fishbone Charting
1.3.3.4.21	Eight-Step Problem Solving
1.3.3.4.24	Gemba Walk—Go and See
1.3.3.4.25	Gap Analysis
1.3.3.4.26	Five Whys
1.3.3.4.27	SIPOC or COPIS
1.3.3.4.28	Pull Signaling
1.3.3.4.29	Affinity Diagrams
1.3.4	Shingo Prize
1.4	Six Sigma
1.4.1	DMAIC
1.4.2	SQC—Statistical Quality Control
1.4.3	SPC—Statistical Process Control
1.4.4	Queuing Theory
1.4.5	DFSS
1.5	QFD—Quality Functional Deployment
1.6	Concept Management
2	Breakthrough Thinking
3	TOC—Theory of Constraints, Bottleneck Analysis, Constraint Analysis
4	Process Reengineering
5	ISO Standards
7	Simulation/Modeling

What Quality Methodology Did the Author Use?: The author chose the quality methodologies listed in Table 5.15.

What Were the Results?: Even though the future state map identified opportunities that would generate a 90 percent reduction in touch points and a 50 percent reduction in cycle time, the limitations of the MES system didn't allow all the future state improvements to be implemented. In the end, the touch points were reduced by 75 percent and the cycle time by about 40 percent.

What Other Quality Methodologies Could Have Been Used?: As we can see from Table 5.14, any number of tools could have been substituted

Table 5.15 Poor Manufacturing Control of Production Activity

Case # 5	Tools Used by the Author
1.3	Lean
1.3.2	Scan
1.3.3	Kaizen Events, Rapid Improvement Events (RIE)
1.3.3.1	PDCA—Plan Do Check Act
1.3.3.2.01	Root Cause Analysis
1.3.3.4	Technical Tools
1.3.3.4.01	Value Stream Mapping
1.3.3.4.02	Seven Wastes
1.3.3.4.03	System Flow Chart
1.3.3.4.04	Swimlane Chart
1.3.3.4.07	Cellular Work Design
1.3.3.4.10	Poka-Yoke (Error Proofing/Mistake Proofing)
1.3.3.4.13	Visual Workplace
1.3.3.4.18	Standard Work
1.3.3.4.19	Brainstorming
1.3.3.4.24	Gemba Walk—Go and See
1.3.3.4.25	Gap Analysis

to complete this project. For example, Six Sigma could have been used to evaluate data transmission and quality errors.

Case #6: Attorney's Office Workflow Inefficiencies

The Organization: This is a government agency that is composed of 80 legal offices. Each office is staffed with a lead attorney and office manager who share the responsibility for running the office, a number of attorneys, and several dozen staff and support employees.

The Process/System: The system comprised the overall workflow of each of the individual offices. Some offices were efficient, but most had extremely poor turnaround time.

The Problem: The problem was that the workflow through the offices was extremely slow with a large number of delays. Customers were extremely dissatisfied with the work.

The Metrics/Goals: The goal was to increase throughput, which was measured in average number of days required to process a case.

Table 5.16 Attorney's Office Workflow Inefficiencies

Case # 6

Characteristics (From Table 4.1)

Quality System	Quality Tool	Quality Philosophy	Acceptance Tool	Technical Tool	Control Tool
		*		*	

Characteristics (From Table 4.2)

Contains Specific Metrics	Contains Implied Metrics	Process/Systems-Oriented	Targets Variability	Targets Process Flow	Top Leadership Involved
	*	*			*

Metrics (From Table 4.3)

Long Term					Short Term				
Customer Satisfaction	Strategic Metrics	Inventory/ Operating Exp	Customer On-time Delivery	Quality Units Produced	Statistical Measures	Cycle Time/ Throughput	Touch Points	Product Lost or Damaged	Not Tied to a Specific Metric
		*				*			

The Problem Characteristics: In Table 5.16 we see how the mapping of Case #6 looks.

Mapping the Problem Characteristics to the Quality Methodology Characteristics: Table 5.17 shows which methodologies could be implemented to help with the problem described in Case #6. The author will only list those that he may reasonably use in this situation.

Table 5.17   Attorney's Office Workflow Inefficiencies

Case # 6	Analyzed Against Tables 4.1, 4.2, and 4.3
1.1	TQM—Total Quality Management
1.1.1	TQM—Deming Version
1.1.2	TQM—Crosby Version
1.1.3	TQM—Juran Version
1.3	Lean
1.3.1	Process, Project, or Event Charter
1.3.2	Scan
1.3.3	Kaizen Events, Rapid Improvement Events (RIE)
1.3.3.1	PDCA—Plan Do Check Act
1.3.3.2	A3 Reporting
1.3.3.2.01	Root Cause Analysis
1.3.3.3	Acceptance Change Management Tools
1.3.3.3.01	Change Acceleration Process (CAP) Model
1.3.3.3.02	Kotter Change Model
1.3.3.3.03	Myers Briggs
1.3.3.3.04	JoHari Window
1.3.3.3.05	Ladder of Inference
1.3.3.3.06	PAPT
1.3.3.3.07	Situational Leadership
1.3.3.4	Technical Tools
1.3.3.4.01	Value Stream Mapping
1.3.3.4.02	Seven Wastes
1.3.3.4.05	Spaghetti Chart
1.3.3.4.06	5S
1.3.3.4.13	Visual Workplace
1.3.3.4.18	Standard Work

(Continued)

Table 5.17 Attorney's Office Workflow Inefficiencies (*Continued*)

Case # 6	Analyzed Against Tables 4.1, 4.2, and 4.3
1.3.3.4.19	Brainstorming
1.3.3.4.20	Fishbone Charting
1.3.3.4.21	Eight-Step Problem Solving
1.3.3.4.24	Gemba Walk—Go and See
1.3.3.4.25	Gap Analysis
1.3.3.4.26	Five Whys
1.3.3.4.28	Pull Signaling
1.3.3.4.29	Affinity Diagrams
1.4	Six Sigma
1.4.1	DMAIC
1.4.2	SQC—Statistical Quality Control
1.4.3	SPC—Statistical Process Control
1.4.4	Queuing Theory
1.4.5	DFSS
1.6	Concept Management
2	Breakthrough Thinking
3	TOC—Theory of Constraints, Bottleneck Analysis, Constraint Analysis
4	Process Reengineering

What Quality Methodology Did the Author Use?: The author chose the quality methodologies listed in Table 5.18.

What Were the Results?: The results varied from one office to another. It depended greatly on willingness of office management. Without their participation, the recommended changes weren't implemented and no improvements were visible. In one example, the author found an employee who had generated a one-year backlog simply by being nonresponsive

Table 5.18 Attorney's Office Workflow Inefficiencies

Case # 6	Tools Used by the Author
1.3.2	Scan
1.3.3.4.24	Gemba Walk—Go and See
3	TOC—Theory of Constraints, Bottleneck Analysis, Constraint Analysis

to letters that were received. The letters were simply thrown in a drawer unopened. However, most of the offices showed a marked improvement. For example, the biggest success was a Dallas office where, two weeks after implementing the recommended improvements, they experienced a tripling of their throughput.

What Other Quality Methodologies Could Have Been Used?:   As we can see from Table 5.17, any number of tools could have been substituted to complete this project. For example, either Lean or Six Sigma could have been used to evaluate workflow.

Case #7: High Quality Defect Rate in Printing Process

The Organization: This is a high-tech manufacturing factory that was experiencing enormously high defect rates. They produced electroluminescence products, like vehicle dashboards or many of the flashing signs and displays that you see in casinos.

The Process/System: The system was the production flow process and the need to identify a way to reduce the error rate in this flow.

The Problem: The defect rate was over 14 percent, and the specific assignment was to reduce this high error rate.

The Metrics/Goals: The primary metric was the defect rate, with a goal of reducing it by 50 percent, but a secondary goal was to reduce inventory.

The Problem Characteristics: In Table 5.19 we see how the mapping of Case #7 looks.

Mapping the Problem Characteristics to the Quality Methodology Characteristics: Table 5.20 shows which methodologies could be implemented to help with the problem described in Case #7. The author will only list those that he may reasonably use in this situation.

What Quality Methodology Did the Author Use?:   The author chose the quality methodologies listed in Table 5.21.

What Were the Results?:   As shown in Figure 5.8, the defect rate was driven from the original 14+ percent to below 2 percent in less than a year, and finished goods inventory was also cut in half during this time.

What Other Quality Methodologies Could Have Been Used?:   As we can see from Table 5.20, any number of tools could have been substituted to complete this project. For example, either Lean or theory of constraints could have been used to evaluate the quality failures.

Table 5.19 High Quality Defect Rate in Printing Process

Case # 7 — Characteristics (From Table 4.1)

Quality System	Quality Tool	Quality Philosophy	Acceptance Tool	Technical Tool	Control Tool
*	*		*	*	*

Characteristics (From Table 4.2)

Contains Specific Metrics	Contains Implied Metrics	Process/Systems-Oriented	Targets Variability	Targets Process Flow	Top Leadership Involved
*		*	*	*	*

Metrics (From Table 4.3)

Long Term					Short Term				
Customer Satisfaction Metrics	Strategic Metrics	Inventory/ Operating Exp	Customer On-time Delivery	Quality Units Produced	Statistical Measures	Cycle Time/ Throughput	Touch Points	Product Lost or Damaged	Not Tied to a Specific Metric
*				*		*		*	

Table 5.20   High Quality Defect Rate in Printing Process

Case # 7	Analyzed against Tables 4.1, 4.2, and 4.3
1	TPS—Toyota Production Systems
1.1	TQM—Total Quality Management
1.1.1	TQM—Deming Version
1.1.2	TQM—Crosby Version
1.1.3	TQM—Juran Version
1.2	TQC—Total Quality Control
1.3	Lean
1.3.1	Process, Project, or Event Charter
1.3.2	Scan
1.3.3	Kaizen Events, Rapid Improvement Events (RIE)
1.3.3.1	PDCA—Plan Do Check Act
1.3.3.2	A3 Reporting
1.3.3.2.01	Root Cause Analysis
1.3.3.4	Technical Tools
1.3.3.4.01	Value Stream Mapping
1.3.3.4.10	Poka-Yoke (Error Proofing/Mistake Proofing)
1.3.3.4.18	Standard Work
1.3.3.4.24	Gemba Walk—Go and See
1.3.3.4.25	Gap Analysis
1.3.4	Shingo Prize
1.4	Six Sigma
1.4.1	DMAIC
1.4.2	SQC—Statistical Quality Control
1.4.3	SPC—Statistical Process Control

Table 5.21   High Quality Defect Rate in Printing Process

Case # 7	Tools Used by the Author
1.2	TQC—Total Quality Control
1.3.3.4.18	Standard Work
1.3.3.4.24	Gemba Walk—Go and See
1.4	Six Sigma
1.4.1	DMAIC
1.4.2	SQC—Statistical Quality Control
1.4.3	SPC—Statistical Process Control

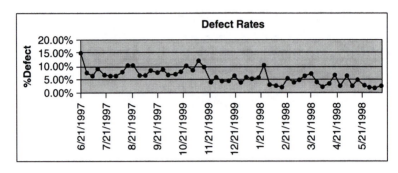

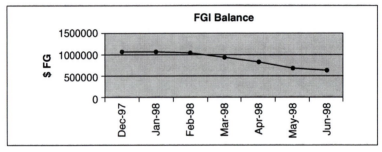

Figure 5.8   Quality transformation.

Summary

This chapter focused on specific problem examples, showing what tools can be used to solve those problems and describing specifically how the author solved the problem. Next we will generalize the quality methodology selection procedure by giving the reader a quality decision tool that they can use.

PART III

Executing the Improvement Process

CHAPTER 6

Identifying the Correct CPI Tool

Professionals built the Titanic—amateurs the Ark.

FRANK PEPPER

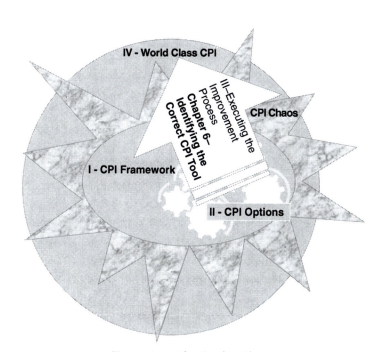

Figure 6.1 The Quality Chaos.

This chapter summarizes the procedures that were used as examples for the seven cases detailed in Chapter 5. Using the definitions supplied in Chapter 4 for each of the classification categories, we look at our problem area (opportunity for improvement) and attempt to classify it using Figure 6.1. Sometimes we don't know enough about the problem so that we feel comfortable classifying it. In that case it is useful to do a scan (Methodology 1.3.2) or develop the project charter (Methodology 1.3.1), after which we will have a much clearer understanding of the effort that we are about to engage in.

Once we have classified the characteristics of the problem/opportunity that we are focused on, we can then compare them to the characteristics of each methodology, as listed in the Tables 4.1, 4.2, and 4.3. As we have seen in Chapter 5, rarely do we find a quality improvement methodology that will work as a stand-alone and fit all the characteristics of our project. Normally, the solution will require a combination of several tools.

Also, it should be noted that a lot of the tools work together. Looking at Figure 6.2 for reference, we see a grouping of tools. For example, by itself, Lean (1.3) is just a methodology for identifying the appropriate change management tool and applying it correctly to the problem. It needs to be complemented with some front-end tools, such as a scan (1.3.2) or a charter (1.3.1), to clearly identify what is being solved. Then the change management/acceptance process needs to be evaluated (1.3.3.3) and, if necessary, a change management tool needs to be applied (1.3.3.3.01 to 1.3.3.3.07), or the project has very little chance of succeeding. Sometimes a reporting/governance tool is also needed (1.3.3.1). Now we are finally ready to identify and select appropriate technical tools (1.3.3.4) focused on the specific problems that we are trying to solve. The list of tools is 1.3.3.4.01 to 1.3.3.4.29. The classifications that were generated in Table 6.1 are then matched to the classifications in Tables 4.1, 4.2, and 4.3 to find the best tools.

This same type of grouping occurs with Six Sigma (1.4). Six Sigma does not often use acceptance tools, but if the situation requires it, they can be applied here (1.3.3.3 to 1.3.3.3.07). Six Sigma has tools that are specific to its practice (1.4.1 to 1.4.5), but it also uses many of the same tools that were identified in the Lean technical tools section (1.3.3.4). For example, Six Sigma uses gap analysis (1.3.3.4.25) and COPIS (1.3.3.4.27). Any of the tools identified as Lean technical tools (1.3.3.4) can be used as technical tools for Six Sigma.

Similarly, TOC (3) to process reengineering (4) can use the same acceptance tool set (1.3.3.3) or technical tool set (1.3.3.4) that was listed under Lean (1.3). These tools are not exclusive to any of these practices.

Table 6.1 Lack of Organizational Strategic Alignment with No Focus on Quality

Case # 6.1	Characteristics (From Table 4.1)	Quality System	Quality Tool	Quality Philosophy	Acceptance Tool	Technical Tool	Control Tool
	Characteristics (From Table 4.2)	Contains Specific Metrics	Contains Implied Metrics	Process/Systems-Oriented	Targets Variability	Targets Process Flow	Top Leadership Involved

Metrics (From Table 4.3)	Long Term				Short Term				
	Customer Satisfaction Metrics	Inventory/ Strategic Operating Exp	Customer On-time Delivery	Quality Units Produced	Statistical Measures	Cycle Time/ Throughput	Touch Points	Product Lost or Damaged	Not Tied to a Specific Metric

Moving forward with our selected quality methodologies, we plan our quality improvement activity based on the procedures utilized by this methodology. Chapter 3 gave a brief definition of each of the methodologies, but this was not intended to be sufficient for anyone to use the tool. Numerous books are available that the reader can use in order to get an in-depth understanding of any of the specific tools listed in Table 6.2.

Table 6.2 Quality Methodology

1	TPS—Toyota Production System
1.1	TQM—Total Quality Management
1.1.1	TQM—Deming Version
1.1.2	TQM—Crosby Version
1.1.3	TQM—Juran Version
1.2	TQC—Total Quality Control
1.3	Lean
1.3.1	Process, Project, or Event Charter
1.3.2	Scan
1.3.3	Kaizen Events, Rapid Improvement Events (RIE)
1.3.3.1	PDCA—Plan Do Check Act
1.3.3.2	A3 Reporting
1.3.3.2.01	Root Cause Analysis
1.3.3.3	Acceptance Change Management Tools
1.3.3.3.01	Change Acceleration Process (CAP) Model
1.3.3.3.02	Kotter Change Model
1.3.3.3.03	Myers Briggs
1.3.3.3.04	JoHari Window
1.3.3.3.05	Ladder of Inference
1.3.3.3.06	PAPT
1.3.3.3.07	Situational Leadership
1.3.3.4	Technical Tools
1.3.3.4.01	Value Stream Mapping
1.3.3.4.02	Seven Wastes
1.3.3.4.03	System Flow Chart
1.3.3.4.04	Swimlane Chart
1.3.3.4.05	Spaghetti Chart
1.3.3.4.06	5S
1.3.3.4.07	Cellular Work Design
1.3.3.4.08	JIT—Just-in-Time

1.3.3.4.09	Agile
1.3.3.4.10	Poka-Yoke (Error Proofing/Mistake Proofing)
1.3.3.4.11	PQ Analysis
1.3.3.4.12	TPM—Total Product (Productive, Preventative or Production) Maintenance
1.3.3.4.13	Visual Workplace
1.3.3.4.14	DFM—Design for Manufacturability
1.3.3.4.15	SMED—Single Minute Exchange of Die, Quick Changeover
1.3.3.4.16	Kanban
1.3.3.4.17	Jidoka
1.3.3.4.18	Standard Work
1.3.3.4.19	Brainstorming
1.3.3.4.20	Fishbone Charting
1.3.3.4.21	Eight-Step Problem Solving
1.3.3.4.22	SWOT—Strengths, Weaknesses, Opportunities, Threats
1.3.3.4.23	VOC—Voice of the Customer
1.3.3.4.24	Gemba Walk—Go and See
1.3.3.4.25	Gap Analysis
1.3.3.4.26	Five Whys
1.3.3.4.27	SIPOC or COPIS
1.3.3.4.28	Pull Signaling
1.3.3.4.29	Affinity Diagrams
1.3.4	Shingo Prize
1.4	Six Sigma
1.4.1	DMAIC
1.4.2	SQC—Statistical Quality Control
1.4.3	SPC—Statistical Process Control
1.4.4	Queuing Theory
1.4.5	DFSS
1.5	QFD—Quality Functional Deployment
1.6	Concept Management
2	Breakthrough Thinking
3	TOC—Theory of Constraints, Bottleneck Analysis, Constraint Analysis
4	Process Reengineering
5	ISO Standards
6	Strategic Mapping
6.1	Hoshin Planning
7	Simulation/Modeling

The Role of Measurement and Reward Systems

Innovation and creativity need to be stimulated. Recently, the author was working for a company that had an elaborate TQM program. However, the TQM program was failing and they couldn't understand why. In reviewing their measurement system, we quickly learned that they were evaluating employee performance based on units per hour efficiency. Bonuses were being paid when employee performance exceeded the standard rates of production. Why would any employee want to spend time implementing changes through TQM if

1. They were being rewarded based on historical rates and historical methodologies
2. There was no reward for implementing the change
3. Changes would, in effect, decrease their productive output on the short term and therefore reduce their bonuses

When change is implemented, the first thing that happens is a drop-off in efficiency. As we see in Figure 6.2, we are working away at a certain level of output, and immediately, as change is introduced, a loss of productive output occurs. Then, through the process of the learning curve, employees slowly become better and better at the new process, eventually achieving a new, higher (hopefully) level of output. Unfortunately, in the short term, efficiency suffers, and so does the paycheck.

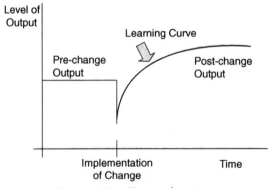

Figure 6.2 Change function.

This company was asking the employees to sacrifice their paycheck in order to implement changes. The company didn't want to lose output, but they still wanted the employees to initiate improvements. They were giving the employees mixed signals, and the signal that motivated the employees the most was the paycheck signal.

To understand this concept completely, here is a "hang on your wall" statement:

A Measurement System Is Not for Management Information; It's for Motivating Employees.

Any measurement system that exists simply for "accounting purposes" or "information gathering purposes" is probably countereffective and is destructive to the company's ability to achieve its goals. A measurement system that is not focused on the goals of the organization is distracting employees away from the goal of the organization. This is because, whether it is true or not, employees believe they are being graded by what they are being measured on. And they will focus their performance in those areas that they are being measured on. So, select your measurement systems wisely; they may motivate the wrong actions.

In the change model we start by operating at a stable level of operation. Then change is implemented. The level of efficiency drops, and a new learning curve kicks in. The growth stage is the most critical stage because this is the time when many changes are dropped. If the growth stage takes too long, the change may be dropped. This is what occurred with Florida Power and Light; with the United States Air Force; and in many JIT, TQM, or process reengineering implementations. Unfortunately, when a change process is dropped during the growth phase, the organization wastes the growth that has occurred, which would have brought them to a newer, higher level of performance. Most U.S. companies don't want the growth stage to take more than a few months, and often, with larger changes, this short time span is impossible.

Eventually, after the growth has leveled out, we start to see a return on the change process. The final phase of the learning curve has kicked in. Finally, we have once again achieved stability, hopefully at a higher level of output.

It matters not if you tried and failed and tried and failed again.
What matters most is if you tried and failed and failed to try again.

Defining the Measurement System

Measuring too much is no more valuable than measuring nothing because it confuses the motivational signals.

So how do we select a measurement system? Some process simplifications can be applied. If we are interested in quality, we should measure defect performance rates, on-time delivery performance, employee change recommendations, and employee turnover. An excellent collection for world class areas of performance measurement is found in the Baldrige criteria. This changes from time to time, so an updated version should be accessed (search the Web for "Baldrige Award"). The basic award criteria are built around

- Quality of products and services
- Comparison of quality results
- Business process and operational and support-service quality improvement
- Supplier quality improvement

Often, false measures are introduced that claim to be something that they are not. For example, ISO 9000 certification or QS certification claim to be a measure of quality performance. They offer a structure upon which a quality system can be built. They offer quality tools, like standard work. But they are not a measure of quality performance. Another example is statistical process control (SPC), which claims to be a quality system. However, the author received a call from a company who said that a consultant had come in and installed SPC about one year ago and they have been collecting data ever since. They were wondering what they should do with all the data. My answer was that they should "throw it out" because SPC is a tool for identifying areas that need continuous, real-time, improvement "during the run" rather than just a data collection system. Other companies have found that SPC is an excellent quality motivational tool. They use their SPC system to measure an area where they are trying to motivate better performance and then never really do anything with the data.

When considering worker or process performance, we need to focus on the critical resource for measurement. Measuring a process performance indicator like cycle time affects many areas, like customer on-time delivery performance, quality, inventory levels, and so on. It is often more meaningful than "quality units produced by employee."

A measurement system should be used to motivate, not as a data collection device.

A quality methodology like Lean focuses on the long-term health of your organization. This means that your Lean enterprise should focus on performance. And improving performance means improving quality, cost, and delivery. To facilitate and motivate this, there are numerous effective measures of Lean performance. The most commonly used operational metrics are focused on

1. Cycle time, which incorporates inventory reductions and capacity increases
2. On-time performance with regard to customer expectations
3. Quality, which is the foundational building block of a satisfied customer base
4. Lead time, which causes buffers of materials and capacity
5. Total cost of operations

But the most important criterion for an effective measurement system in any IT environment is that it focus on motivating the correct response from the employee base. The measure that will fit your organization the best depends on

The goals of the organization
The expectations of the customer
The response that employees or suppliers will have to the measure
The accessibility and reliability of the measure

By evaluating these criteria, you identify as few measures as possible that will drive everyone's response toward optimizing that measure. You should never base your measurement system on tradition.

The Role and Purpose of Control Systems

Too many organizations react to failures or problems by adding control systems, like statistical process control (SPC), which, in and of itself, is a very good system when applied as a performance enhancement tool and is used to measure process performance "as it happens." But as a control system "after the fact," it has minimal effect on quality output. It identifies a varying number of areas where errors occurred, but it does little to aid in identifying solutions.

Inappropriately applied control systems are the enemy of an efficient production environment.

Control systems should not be arbitrarily applied. Control systems

▲ Add steps to the process
▲ Increase the opportunities for failure since there are now more steps in the process
▲ Increase the overall cycle time
▲ Misdirect employees on what's important in achieving overall goals
▲ Waste resources (time, floor space, etc.)
▲ Waste capacity
▲ And most importantly, they move the error to somewhere else in the process rather than fixing or eliminating the error

If there is a problem (opportunity) in the process, it should be resolved by rethinking the system, not by adding another step into the system, which is another opportunity for failure to occur. The tools discussed in this book facilitate correcting the system without adding complication to it. In general, a control system is an indicator that the overall process is not working correctly.

If you're on a path with no obstacles, you're going nowhere.

Summary

This chapter summarized the procedure that was used in the examples of the seven cases in Chapter 5 using the definitions supplied in Chapter 4 for each of the classification categories. This chapter also discussed metrics and control systems. Without meaningful measures, you will achieve your goal: NO MEANINGFUL RESULTS. And with incorrect measures, you will receive incorrect results. The selection and proper implementation of a measurement system are critical to successful quality CPI process improvements.

People respond to what is INSPECTED, not what is EXPECTED.

CHAPTER 7

The Role of Governance, Performance Reviews, Cascading, and Communication

Obstacles are what we see when we take our eyes off of the goal.

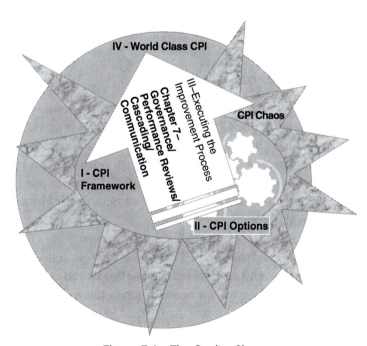

Figure 7.1 The Quality Chaos.

This chapter discusses the sustainment and on-going stability of a quality-based CPI system. Without a sustainment structure in place, it is impossible to maintain a valid quality-based "continuous" process improvement system, no matter what methodology is chosen. Effective sustainment requires the following components:

- A governance system—The purpose of the governance system is to keep the strategic CPI vision in the forefront of everyone's mind. It's a status check of progress toward goals.
- Regular performance reviews—The performance reviews are the formal reporting and feedback mechanisms that check progress and report results.
- A methodology for cascading—Cascading is about sharing the enthusiasm and commitment of CPI and involving everyone throughout the company in the CPI strategy. How do we get them to take ownership in the process?
- A communications plan—The communications plan is about sharing information updates with the entire organization.

We will discuss each of these separately, and then we will integrate them together into a sustainment methodology for CPI.

A Governance System

The "governance system" is the fancy term we use for monitoring and tracking the progress of our CPI systems. This includes progress reports, corrective action, and feedback mechanisms. As we can see in Figure 7.2, an effective governance system is built on the foundation of an effective organizational strategy. It is the link, the centerpiece, between this strategy and the CPI/quality methodologies. It is what sustains the CPI process and keeps it aligned with the strategy.

Strategic CPI governance is the methodology for systematically making decisions based on the actual results. These results are generated using the metrics of both internal and external influences (threats and opportunities) that impact strategic performance as we progress toward implementing specific operational tasks.

The performance reviews should become the centerpiece of CPI/strategic decision making and execution. CPI/strategic decision making is

Figure 7.2   Governance.

never static and will always be impacted by the constant flow of new information. The key to successful CPI/strategy governance is to be flexible and responsive to change.

CPI/strategic governance has the following critical success drivers:

- To assure that leadership plans have not been sidetracked by extraneous issues—**Accountability**.
- CPI/strategic analysis has changed how operational decisions are made—**Focus**.
- The work culture has been transformed into a CPI culture so it is easier to learn, share, and grow in strategic and operational knowledge—**Teamwork**.
- There is an increase in innovative CPI/improvement opportunities and "out of the box" creative thinking—**Results**.

In particular, when we're looking at governance, we're looking for a structure that provides the following:

- A methodology for executing strategic plans (see Quality Methodology 6 defined in Chapter 3).
- A feedback mechanism that will monitor, adjust, and align the strategy map, thereby keeping it relevant to business trends and changes.
- A mechanism that will facilitate the optimization of resources. Resources are always limited, and all the desired changes cannot be implemented at the same time, so prioritizations and resource allocations will need to be made.

▲ Governance should make sure the crosschecks are in place that will keep the strategy map as the centerpiece of all decision-making processes.

▲ Accountability needs to be tracked to assure that commitments are being met.

▲ Workforce involvement needs to be monitored to assure that strategic execution occurs.

When a CPI governance process is put into practice, there will be a concerted effort within the organization to focus on executing high-impact, goal-focused improvements. This is facilitated by CPI execution feedback (information roll-up), which is consistently coming from the employees as they implement strategic change.

The strategy is the main document to communicate strategic plans, and the CPI plan is the execution of those plans throughout the organization. Performance can be assessed using these reporting documents and tools during regularly scheduled reviews:

▲ Scorecard Report—High-level summary of the strategy map's objectives and corresponding CPI action taken.

▲ Dashboard Report—Summary of strategic operational metrics used to monitor the progress of the strategic objectives.

▲ A3 Report—Review of a specific CPI improvement activity and the steps being implemented to ensure the strategic objective is executed correctly. The A3 report is also used to highlight results accomplished and to indicate execution problem areas that require additional help.

The governance process is a communication tool used to depict the organization's priorities, goals, and objectives. This will give organizational leadership and the workforce the opportunity to continuously discuss strategic/CPI organizational direction, coordinate the creation of deliverables, instill accountability, and reinforce the need to collaborate efforts. This will drive us to achieve the highest possible joint success. When the strategy and CPI plans are used as a communication tool, the workforce will be equipped to make strategic execution their responsibility.

Regular Performance Reviews

Performance reviews are the glue that makes sustainability work. Performance reviews build accountability into the CPI process by creating regular checkpoints. Successfully facilitated CPI performance reviews will collaboratively achieve the following outcomes:

▲ Identify new insights by analyzing performance results
▲ Create opportunities for integrated team problem solving
▲ Devise and evaluate new improvement initiatives and tasks
▲ Enhance CPI performance management
▲ Assess, on a recurring basis, external and internal environments and events that impact execution of strategic CPI initiatives

To achieve the desired outcomes, leadership will be tasked with keeping the review briefing focused on the relevancy of the CPI initiatives and their strategic objectives. These meetings should create an environment that promotes candor in sharing experiences and intellectual capital. This fosters a common purpose, encouraging employees to think beyond their own job responsibilities. Management needs to allow discussions into operational or tactical issues.

Effective review briefings are dependent upon management's clear understanding and accepting their responsibilities with regard to achieving their strategic objective's desired outcomes. These roles include

▲ Developing and coordinating strategic action plans
▲ Establishing a team of subject matter experts (SMEs) to promote cross-functional integration of activities that support the objective's planning, execution, and performance
▲ Identifying appropriate and meaningful performance measures with short- and long-term stretch targets
▲ Applying analytics to the measurement data to identify trends, operational barriers, and effectiveness and efficiency improvements, thereby removing obstacles that will inhibit progress
▲ Delivering a performance assessment report for review briefings to cultivate a sense of ownership

Following are some additional guidelines for use in assessing and reporting on each CPI's strategic purpose and any obstacles, action plans, or milestones encountered. The reporting needs to stay focused on the most critical issues that are of interest for the briefing attendees. This approach allows attendees to receive and analyze operational feedback and solve strategic problems.

The following four points should be addressed during any performance review:

1. **Problem/Issue**—What is the fundamental problem/issue that is being addressed? What needs to be solved or changed?
2. **Analysis**—What are the root causes of the problem/issue? What is the impact on the objective? What can be done that will improve performance and eliminate operational barriers?
3. **Action Plan**—What are the specific action plans and initiatives that are being executed? Are the right process improvement activities being selected?
4. **Milestones**—What are the specific milestones that have been established to monitor the action plan's on-going progress?

Performance review results would include

▲ Encouraging continual feedback and review processes to monitor and improve CPI execution
▲ Facilitating anticipatory management, which focuses on creating operational change and producing optimal strategic value
▲ Analyzing obstacles and risks to be overcome
▲ Focusing on the reasons for performance gaps between actual results and operational stretch targets
▲ Unifying leadership to drive strategic change

The heart of CPI execution is the quality of the performance reviews, which bring together the on-going quality change processes. To execute this process, several reporting tools are available. Three dominant measurement reporting tools have proven effective in measuring CPI performance:

1. Dashboard
2. Balanced scorecard
3. Scorecard

The dashboard is a tool that looks at operational measures of performance. For example, if we are measuring production quality, the dashboard would be a matrix table that highlights successes and failures. Looking at Table 7.1 we see an example of a quality dashboard.

In addition, the dashboard would have highlights—for example, the box labeled Business Implementations with Priority 1 Issues Taking

Table 7.1 Quality Dashboard Example

Functional Area	Internal Failures	Customer Complaint First Calls	Customer Complaint Second Calls	Priority 1 Issues Taking More Than Two Days	Priority 2 Issues Taking More Than Two Weeks	Complex On-going Issues Not Resolved
Major Process Failure—Total Product Rejection	12	3	1	2	3	2
Medium Process Failure	7	2	2	3	2	2
Minor Process Failure—Adjustment Was Made and Customer Is Satisfied	5	4	2	4	1	3
Customer Caused Failure	9	3	1	3	1	2

More Than 2 Days is highlighted because its value is higher than the allowable range.

The balanced scorecard is a high-level, organization-wide tool balancing both financial and operational metrics and aligning them with the enterprise vision/strategy. To utilize the balanced scorecard, we evaluate performance in four dimensions, each time defining the CPI and strategic alignment:

▲ Financial perspective—Corporate ROI
▲ Customer perspective—Partnering with your customer to enable success

204 | Part III Executing the Improvement Process

Table 7.2 Quality Scorecard Example

Functional Area	Last Six Months	Last Month	Current Month	Projected Next Month	Projected Next Six Months	Need Executive Support
Quality Initiative A	Red	Red	Yellow	Yellow	Green	Green
Quality Initiative B	Green	Green	Green	Green	Green	Green
Quality Initiative C	Red	Red	Red	Red	Red	Red
Quality Initiative D	Green	Green	Green	Green	Green	Green

▲ Internal process perspective—Develop and maintain an environment that meets CPI and strategic needs

▲ Innovation and learning perspective—Foster an environment that attracts and retains employees and encourages creativity

Identifying the four dimensions is the easy part. Defining the strategic metrics that will support each of these dimensions is the challenge. However, once defined, the dashboard offers a quick and consistent tool to evaluate organizational performance.

The scorecard is different from the balance scorecard, but it can also have many of the same elements of information. The scorecard is a high-level performance report. Unlike the dashboard, which gets into the details of each activity, the scorecard is more of a strategic reporting tool that spotlights progress in each of the strategic areas (red for bad, yellow for struggling, and green for good). Then, if there is an area of interest, a drill-down to a dashboard can occur. Table 7.2 is a brief example of a scorecard.

The scorecard offers a quick visual representation of the organization's performance. Management and employees can quickly see which CPI focus areas are on track and which are in trouble.

A Methodology for Cascading

Cascading is the sharing of high-level corporate plans and initiatives with the lower levels of the organization so that they can identify CPI initiatives that align with the corporate strategies. The goals and purpose behind cascading are the following:

- Execute the strategic priorities through CPI quality methodologies on the front lines
- Align and validate the quality initiatives against the organizational strategy
- Promote workforce feedback and continuous dialogue
- Focus constrained resources on the right strategic tasks and CPI initiatives at the right time and in the right place

Effective quality strategy execution requires the cascading of corporate strategy down to the lower levels of the organization. It ensures the headquarters-level plans are translated into the plans of the various lower-level operating and support units. It includes executing strategic CPI initiatives that deliver on the strategic objectives. It aligns employees' competency development plans and their personal goals with the higher-level corporate strategic plans. It creates bridges between an organization's strategy, processes, systems, CPI initiatives, and, most importantly, its people. In this way, cascading will ignite breakthrough results through effective and efficient strategy CPI execution. Cascading structures a process of focus and coordination across even the most complex organizations.

Strategy cascading is accomplished with a communication plan that shares strategic and quality CPI information through the various levels of the organization. This cascaded information should be aligned with the higher-level strategy while reflecting the unique operations of the lower-level units. It promotes local ownership of their strategic CPI plan. Cascading is also accomplished by forming cross-functional SME work teams. These cross-functional meetings will

- Find new insights for analyzing quality performance results
- Identify quality problem areas, performance gaps, and opportunities for improvement
- Engage in integrated-team quality problem solving
- Devise and evaluate quality action plans
- Enhance quality process improvements
- Access external and internal environments and events that will impact execution of the CPI quality strategy
- Provide evidence about the validity of the assumptions underlying the corporate quality CPI strategy

Without active operational quality strategy cascading and integration, organizations face the risk of establishing competing strategic priorities. This can result in promoting initiative overload and wasting limited resources on redundant tasks. When an organization is aligned around common critical quality strategic priorities, the operational value of meeting customer needs will be greater than the sum of its individual isolated operational units.

Strategic CPI quality management and execution should not be looked at as a separate management responsibility. For CPI to create organizational change, it needs to be integrated with each of the key operational management processes. Without this, resources will be wasted in a duplication of effort. The resource areas that need consideration include

▲ Leadership Management—This includes the reporting, oversight, and operational structures within the organization. These structures need to be continually integrated with the corporate objectives in order to maximize effect on the organizational strategic quality priorities.

▲ Resource Management—This consists of the reporting, management, and allocation of resources (dollars and people). When this is integrated with strategic quality planning, the right resources will be assigned to the right strategic priorities and goals to maximize the return on these investments.

▲ Process Management—This consists of the various quality methodologies or management tools that are available to monitor operational efficiencies and effectiveness. This ensures current processes and procedures are maximizing their intended purposes. It generates operational improvements and the efficient use of limited resources.

▲ Quality and CPI Strategy Management—This consists of the development, governance, communication, and cascading of CPI plans within the organization.

A Communications Plan

A communications plan defines the process that will be used to share strategic CPI and quality initiative information throughout the organization. The goals of the communications plan are:

▲ Communicate the organization's CPI quality strategic plan and corresponding performance results to all internal and selected external audiences

- Advance the CPI quality culture by aligning it with the understanding of every employee
- Energize employees by aligning their work to applicable critical strategic quality objectives
- Empower employees to focus on the right tasks and to do them correctly, thereby realizing major performance improvements
- Motivate employees to understand they are the frontline in successful quality strategy execution

A critical success factor in implementing a quality strategy for any organization is that the right message is being shared, understood, and believed in by the right people. Organizations need a formal program to communicate strategic quality program initiatives. If people do not understand the organization's strategic quality plan, it is the same as the organization having no quality plan at all. Some additional goals of an effective quality strategy communication plan are

- To educate and motivate all employees on the quality strategy processes and its operational value
- To develop a working understanding of the quality strategy so that the workforce knows how specifics of the organization's plan will impact their individual job responsibilities
- To create a proactive commitment and momentum among the employees, which will generate support for the execution of the quality plan
- To generate feedback from the employees on how to best accomplish the intended purpose of selected quality objectives

Effective communication to the workforce about quality strategy and CPI initiatives is vital if the entire workforce is to have the opportunity to contribute to the quality improvement process. This involves an understanding of the organization's quality strategy and of the scorecard. People who are knowledgeable and inspired by their quality plan have the potential to improve the alignment of their work with strategic quality priorities. Then the workforce will be in a better position to identify efficient and effective ways to support the intended outcomes. Organizations will fail in their quality strategy execution if employees do not understand their organization's higher-level strategy and do not know how they can contribute to its success.

The desired communication plan outcomes include

▲ Creating unity and momentum among the entire workforce around a common purpose and strategic quality destination
▲ Demonstrating senior management's commitment and passion about quality
▲ Promoting cultural change, making it easier to learn, share, and grow in knowledge together
▲ Encouraging innovative and "out-of-the-box" creative thinking among all employees, thereby hoping to find new ways to achieve the quality strategy
▲ Facilitating strategic understanding, acceptance, execution, and ultimately, the realization of breakthrough results

> Successful communication is dependent on the right messages being understood and believed by the right people.
>
> PETER DRUCKER

Summary

This chapter discussed the sustainment and on-going stability of a quality-based CPI system. Sustainment puts a structure in place, allowing the continuous growth and development of a quality culture throughout the organization. This chapter mentioned that effective sustainment requires the following components:

▲ A governance system
▲ Regular performance reviews
▲ A methodology for cascading
▲ A communications plan

It's not enough to have a good strategic quality CPI plan. The plan won't take hold in your organizations without a plan for sustaining the process over the long term. This chapter discussed the keys for making it work.

PART IV

World Class CPI

CHAPTER 8

"Moving the Needle"

Today we do the impossible. Tomorrow we create miracles.

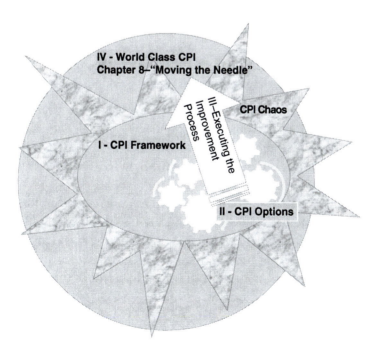

Figure 8.1 The Quality Chaos.

At this point, we know the following:

▲ What quality methodology options are available
▲ The characteristics of each of the quality methodologies
▲ How to evaluate the characteristics of our quality opportunity
▲ How to blend the characteristics of our problem with those of the quality methodologies
▲ How to select the appropriate methodology(ies) to address your problem
▲ The role of metrics in the CPI process
▲ How to sustain an on-going culture of continuous quality improvement

Now we need to take these principles and "move the needle." That means we need to take our current collection of quality concerns and benchmark their current position (measure the current level of performance). Then we need to "move the needle" on those measures to a desired level of performance.

Let's take a brief industry perspective on how best to move forward. Here are just a few of the author's experiences, by industry. The author is not going to give every example of every industry. That would be a book by itself. But these examples should be sufficient to start your thought juices flowing.

Medical industry   In the medical industry, the author has experienced the following key areas of concern:

▲ Extensive, poorly managed, and excessive inventories. Inventories are staged at various locations throughout the hospital, with numerous duplications, and the staff still experience shortages because the right materials are not at the right place at the right time. Metrics: Inventory levels.
▲ Poor scheduling of the most expensive resource, the operating room (OR). The OR is scheduled in blocks, and these blocks are sold to doctors. The scheduling should be done by procedure, and as much of the functional activity as possible should be performed pre- and post-op. The author has experienced triple OR initializations through proper scheduling. Metrics: Capacity utilization.
▲ Poor scheduling of the second most expensive resource, the emergency room (ER). A lot of the problem in the ER is related to elements outside of the ER, such as excessive lab processing lead times, which keeps

the ER bed tied up. Another problem is the bed scheduling for the rest of the hospital, so that identifying bed availability becomes a major effort. Improving the processing times for the ER support organizations allows a more rapid turnover of the ER beds. Metrics: Lab or bed availability processing time.

▲ Recording errors are prevalent in many areas. For example, lab labeling and analysis errors are far too common, causing the samples to be retaken and causing delays in delivering the badly needed results. Metrics: Quality error rate and processing cycle time.

▲ The processing of records for many hospitals is extremely slow, causing delays in admitting and in the emergency room. Metrics: In-processing cycle time.

High-tech manufacturing industry In the high-tech manufacturing industry, the author has experienced the following key areas of concern:

▲ Shop floor control is out of control in most high-tech manufacturers. New pressures require them to shift from a make-to-stock to a make-to-order environment, and they aren't ready for the transition because their shop floor tracking mechanisms are inadequate. They have tried to do a lot of this through separate manual or IT patch systems, but these additional systems have just complicated the process. In addition, they need compliance tracking on all the products they produce, which requires a detailed event management system that tracks country of origin and product content. So they have to rebuild these systems. Metrics: Cycle time, capacity, and IT touch points.

▲ Inventory control is out of control. Inventory is staged as several points along the supply chain in an attempt to have inventory ready for sale, but far too often it's at the wrong points and doesn't get sold. The carrying cost of this inventory, and the corresponding obsolescence cost, is killing the bottom line. Metrics: Inventory turns, inventory obsolescence costs, process cycle times.

▲ Long product delivery lead times. The short product life cycle of the product means that if the product isn't available when the customer is looking for it, then it won't get sold. Metrics: Logistics cycle time.

▲ Processing orders often takes longer than it takes to build the product. This drives customers crazy. Metrics: Customer satisfaction and order processing cycle time.

▲ Digital supply chain capabilities are becoming big in numerous industries. This involves concerns about response time, security, contract development and renewals, and so on. Metrics: SCM cycle time, response time, number of touch points, repeat business.

Distribution and logistics industry   In the distribution and logistics industry, the author has experienced the following key areas of concern:

▲ Warehouse utilization and optimization is a big problem. The author can always find an area in the warehouse where product has been dumped and has been sitting there for many years. And often, this product was express-shipped in, incurring high shipping costs. Now it is unused and obsolete. This results from a combination of poor product tracking in the order process, shipping, and receiving. Metrics: Inventory obsolescence costs, on-time shipments.

▲ Inventory is staged and poorly tracked throughout the logistics process. Where a shipment is and how it is moving through the system is a big black hole. Event management is needed to track the movement of products. Metrics: Customer on-time delivery percent, inventory turns, logistics cycle time.

Oil and gas industry   In the oil and gas industry, the author has experienced the following key areas of concern:

▲ Inventory management is a problem. Inventory is often lost in the supply chain for no good reason. Inventory stockpiles are excessively expensive and have high carrying costs. Metrics: Inventory turns.

▲ Inventory movement and tracking raises its ugly head when shipments occur but the product is "lost to the system" until it arrives at the customer location. Event management, which would track the movement of inventory through the supply chain, is critical. Metrics: Cycle time, touch points.

▲ Margin optimization becomes a problem when competitive pressures are applied. The industry has become lazy in optimizing its processes because it is able to easily adjust the consumer price. Metrics: Margin, operating costs.

Retail industry   In the retail industry, the author has experienced the following key areas of concern (from here on I will not repeat items that have already been listed—for example, inventory optimization is a problem for almost everyone):

▲   The biggest recurring problem with retail is how to have the right product on the shelf when the customer comes for it. It's been the big differentiator among the big retailers because a customer prefers to have as few stops as possible, and if they can get everything in one stop (of course, assuming it's on the shelf when they come), then they will tend to return to that same retailer. The winner tends to be the retailer with the best logistics tools. The goal is to start the restocking process as soon as an item is pulled off the shelf. Metrics: Replenishment lead time/cycle time, percentage of "not on shelf" items, customer satisfaction levels.

▲   New location setups are a problem for franchises, car dealerships, and so on. The lead time required to set up a new business can be excessive. Metric: Cycle time to establish a new business location.

Hospitality industry   In the hospitality industry, the author has experienced the following key area of concern:

▲   The goal is repeat customers. Thus, the focus is on identifying the customers who would be repeat users and making sure that their repeatability requirements are met. Metrics: Customer satisfaction, service down-time.

Military   The military is a combination of many of the other industries already mentioned. In the military, there is a medical organization, which has many of the same problems already listed. There is also a maintenance organization, which has the same inventory and planning/scheduling issues already listed. There is a support function, which includes construction and logistics issues. And there is an operations function. In this section I will only list one item that is unique to the military environment. In the military, the author has experienced the following key area of concern:

▲   The military is plagued with "short-sightedness." What I mean by that is that they build high-level strategic plans, but the commanders who actually run the various bases or depots rarely look "beyond their watch." A commander is in a position for about two years, and he/she wants to have an impact during that time, but beyond that is someone else's problem. In addition, there's the attitude that the next guy will change everything anyway, so don't think beyond the two years.

This destroys the effectiveness of long-term initiatives, especially in the area of quality or process improvement. Metrics: Long-term goal achievement/performance, dashboards, scorecards.

Front office   In the front office, the author has experienced the following key area of concern:

⚠ The front office is plagued with processing delays, primarily because of a large number of control points and check points that have been instituted. Often, the IT systems make it worse rather than reducing cycle time. It's not unusual to find that it takes longer to process a customer order than it takes to build the product. And that simply does not make sense. Right now I am working with one of the world's largest high-tech manufacturers who takes over a week from the receipt of the configuration request to create a quote. Their competition does it in less than a day. Metrics: Processing cycle time, customer complaints with respect to ease of doing business.

Banking industry   In the banking industry, the author has experienced the following key area of concern:

⚠ What I am repeatedly hearing are concerns around the ease of doing business. Online banking capabilities need to be best in class. Transactions like new account setups need to be less painful for the customer. Customer convenience in all transactions is critical. Metric: Transaction cycle time.

Summary

The goal of this chapter was to show the reader the wealth of quality improvement opportunities that exist. There is no limit to the number of ways you can "move the needle" on quality performance. The examples listed in this chapter go beyond the traditional product quality perspective that most people have. The purpose of this chapter was to show quality opportunities in some of the more nontraditional areas. Now it is up to the reader to find opportunities within their own organization where they can "move the needle."

Obstacles are what we see when we take our eyes off the goal.

CHAPTER 9

Summary and Conclusions

And as we let our own light shine, we unconsciously give other people permission to do the same.

NELSON MANDELA

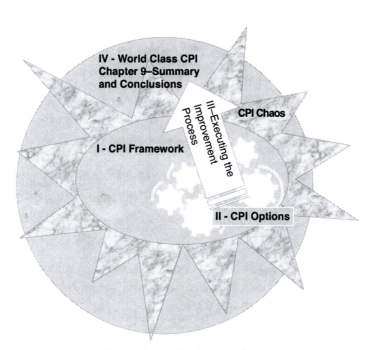

Figure 9.1 The Quality Chaos.

We've made it to the end. Now the real work begins—the work of changing your organization. As we've worked our way through this book, we covered the following topics:

▲   Why is change management necessary and what is CPI?
▲   What quality methodologies are relevant and popular?
▲   What characteristics are relevant when it comes to quality methodologies?
▲   How do these characteristics match with each of the quality methodologies?
▲   How are quality opportunities identified?
▲   How are characteristics of our quality opportunity identified?
▲   How do we match the characteristics of our problem with quality characteristics and them map them to those of the quality methodologies?
▲   How does the selection process of the identified methodology(ies) address our problem?
▲   How do metrics motivate the CPI process?
▲   How does CPI sustainment work? How do utilizing tools like governance, performance reviews, cascading, and a communication program sustain an on-going culture of continuous quality improvement?
▲   How do we "move the needle" when it comes to identifying opportunities within our company?
▲   How do we execute CPI?

> You don't think your way to change. You have to work your way to change.

The last message that I want to reemphasize and leave with you is that none of this will work without two key elements:

1. Top management commitment
2. Cultural transformation

Without top management, the emphasis on a CPI process will come and go without any minor leadership changes. Get the top involved.

The organizational cultural transformation is critical. As mentioned before, no matter how good a system or a change is, if we don't have organizational support (cultural commitment), it will fail. Conversely, no matter how bad a system or change is, if the cultural commitment exists (employee support), it will be a success.

With that said, let me send you on your way with one of my favorite Einstein quotes:

> The definition of insanity is continuing to do the same things and expecting different results.
>
> ALBERT EINSTEIN

INDEX

CPSIA information can be obtained
at www.ICGtesting.com
Printed in the USA
LVOW10*0403300118

564469LV00002B/2/P

9 780071 767187